Risk Management for Water Professionals

This book presents a comprehensive view of the different theories of risk management in water, drawing on recent studies that serve to inform the way that practitioners consider their own risk practice.

While it is commonplace to see risk described in technical and engineering terms when discussing water, this book argues that this is a flawed practice that results in poor decision-making, particularly where water intersects with social elements and the community. Challenging these traditionally held notions of risk, this book introduces the psychological and sociological underpinnings to water risk decisions. Using these, it argues for a broader view of risk-based thinking and proposes a number of evidence-based actions for policymakers to directly implement. Drawing on primary research conducted with water professionals across a variety of roles, this book highlights how the effect of psychological inputs, such as dread and reputation, can create barriers to implementing novel water solutions or projects. Through understanding the biases covered in this book, water practitioners can work to support processes that seek to encourage new and innovative methods in water management.

This book will be of great interest to professionals working in water management, including those in government roles, planning departments and consultancies. It is also a great reference for students of both water resource management and risk studies more generally.

Anna Kosovac is a lecturer in the Faculty of Arts at the University of Melbourne, Australia. She holds a PhD in Risk Perceptions and Decision-Making in the Water Public Sector and has a decade of experience working in the public sector in water management.

Routledge Focus on Environment and Sustainability

Food Cultures and Geographical Indications in Norway
Atle Wehn Hegnes

Sustainability and the Philosophy of Science
Jeffry L. Ramsey

Food Cooperatives in Turkey
Building Alternative Food Networks
Özlem Öz and Zühre Aksoy

The Economics of Estuary Restoration in South Africa
Douglas J. Crookes

Urban Resilience and Climate Change in the MENA Region
Nuha Eltinay and Charles Egbu

Global Forest Visualization
From Green Marbles to Storyworlds
Lynda Olman and Birgit Schneider

Sustainable Marketing and the Circular Economy in Poland
Key Concepts and Strategies
Anita Proszowska, Ewa Prymon-Ryś, Anna Kondak, Aleksandra Wilk and Anna Dubel

Risk Management for Water Professionals
Technical, Psychological and Sociological Underpinnings
Anna Kosovac

For more information about this series, please visit: www.routledge.com/Routledge-Focus-on-Environment-and-Sustainability/book-series/RFES

Risk Management for Water Professionals

Technical, Psychological and Sociological Underpinnings

Anna Kosovac

First published 2024
by Routledge
4 Park Square, Milton Park, Abingdon, Oxon OX14 4RN

and by Routledge
605 Third Avenue, New York, NY 10158

Routledge is an imprint of the Taylor & Francis Group, an informa business

British Library Cataloguing-in-Publication Data
A catalogue record for this book is available from the British Library

ISBN: 978-1-032-55659-8 (hbk)
ISBN: 978-1-032-55870-7 (pbk)
ISBN: 978-1-003-43264-7 (ebk)

DOI: 10.4324/9781003432647

Typeset in Times New Roman
by codeMantra

Contents

1 Risk and Water – Why Should We Care?

Risk is not an area that rouses significant amounts of excitement upon its mention, nor is it a field that veers too heavily from its quantified past. Chances are, in mentioning risk to you, you are likely to recall imaginings of a document with a table presided by likelihood and consequence numerical justifications. Although this is commonplace in organisations such as Environmental Protection Authorities and Water Utilities worldwide, it is not without its flaws. And its flaws are many. The point is that the person who undertakes the assessment is, by most accounts, human. This person will carry their own experiences, biases and socially driven values that underpin the assessment decisions they make. And yet, we presume that allocating a number to a risk calculation will render these impacts as obsolete, that by calling upon the gods of quantification, we can presuppose objectivity. This objectivity is seen as the ideal goal of management, particularly that of public sector decision-making. When spending public money, practitioners want to seem like they are making transparent decisions that seem obvious when others (read: auditors) peruse the documents. It is understandable, but these numbers often create a false sense of security. I will delve further into why this is the case within this book. I will further explore notions of risk, and how they have changed over the years. In particular, how we as a society have shifted towards risk-based thinking and how this has in turn given us a shared sense of control over our environment, but equally, an obsessive tendency to risk manage all our actions.

An argument that is reiterated continuously in this book is that risk is not solely the domain of physical and biological sciences, and to conduct assessments genuinely, we must cognise the role that social structures, psychology and norms play in shaping them.

The Water Issue

The notion of water management practices is heavily underpinned by risk understandings. This is due to the constant supply/demand issues of water resources that stem from a range of factors, including climate change, political

DOI: 10.4324/9781003432647-1

upheaval, urbanisation and economic conditions. Water management emerged from the field of engineering, stemming back from as far as Ancient Greece in its development of Aqueducts to transfer water. As a result, water risk management practice has been developed through this technical-based prism. The word 'risk' itself originated from the Greek word 'rhiza', which is in reference to the hazards of sailing near a cliff (see Figure 1.1). The dangers of water underpins the foundations of 'rhiza' and thus has always been intertwined with risk imaginings. There was the risk of flooding, drought, and poor water quality to consider.

The Asipu people in the Tigris-Euphrates valley about 3200 BC were some of the oldest risk assessors known to date and would conduct risk analyses for those around them. They could be consulted for risky ventures, new construction sites and marriages to identify, analyse and assess the risk with the best available data being drawn from signs from Gods, which they alone could interpret (Covello & Mumpower, 1985). Today, we have moved away from the spiritualistic nature of risk assessment, but still retain the basic structure of the Asipu people's approach. Pascal introduced elements of probability into risk assessments in 1657 (although there have been arguments made that probability extended well before Pascal, see, for example, Franklin (2015)). Despite many centuries of probability theorems to analyse risk, the belief that moral transgression, will lead to subsequent punishment from God, still persisted throughout much of the 17th–19th centuries. It was not until the early 20th century that we begin to see a move away from spiritual and religious risk justifications towards increasingly sophisticated quantification models. This came with the shifting in responsibility and blame of risks, from higher spiritual powers to governments. Within the field of water, we begin to see risk analyses in the form of cost-benefit assessments, particularly with reference to lives lost. The Snowy Hydro Scheme constructed in 1949 and completed in 1974 in South-Eastern Australia was assessed on the basis of an 'acceptable' risk of one life lost per mile constructed. With a rate of 0.9 lives per mile, it was considered to be successful, but to what end? Our understandings of these risks have not shifted substantially, when considering the way that public health risk assessments are taken which recognise some harm from certain approaches, such as fluoride in water or chlorination, but ultimately deem it within the figures acceptable for safety.

Governments have thus always played a role in mitigating risk to its citizens, and water plays a central part of this responsibility. The delivery of water services come with inherent risks attached to them that are decided upon by decision-makers in the public sector. As clean water and wastewater exist as a core part of livelihoods, the added complexity of a lack of alternatives (i.e. you cannot substitute other items in the place of clean water) makes water and wastewater a particularly indispensable resource, which may in itself decrease the risk that decision-makers are willing to accept.

Figure 1.1 'The Shipwreck' (1772) Claude-Joseph Vernet.

Episodes of dire outcomes in a lack of safe water and wastewater services not only are a common occurrence in Global South locations but also have been seen in the Global North. For example, the much-publicised case of the City of Flint in Michigan, United States. In 25 April 2014, as a result of expected demand shortages, and (as some The Natural Resources Defense Council (2018) argue) a cost-saving move, the City of Flint decided to shift their source water from the Detroit Water and Sewerage Department to treated water from the Flint River. This was meant to exist as an interim solution, while the new pipeline to nearby Lake Huron was built, which was to be the future water supply for Flint. There were issues with the water immediately, as there were high levels of trihalomethanes within the water, an organic molecule linked to numerous health concerns. In addition to this, the change of water had a notable impact on how the water reacted with existing lead distribution pipes in the system. The previous water supply ensured that phosphate ions were added to the water, reducing the chance of lead from the pipes entering the supply. Authorities did not treat the Flint River water with phosphate ions, leading to lead ions entering the water system and into homes. This led to levels of lead in water supply reaching 1,000 times the action level set by the Environmental Protection Agency in the United States. The resulting health effects of elevated blood lead levels have been numerous and catastrophic (Craft-Blacksheare, 2017), ranging from seizures, cognitive decline, hypertension, cardiovascular effects to ongoing ill mental health (Cuthbertson, Newkirk, Ilardo, Loveridge, & Skidmore, 2016; Hanna-Attisha, LaChance, Sadler, & Champney Schnepp, 2015). This is a dire failing of effective risk management, and risk considerations. Poor water quality can have ongoing ill outcomes on populations for decades to come.

When considering risk, it is rarely housed in a social and cultural vacuum. This is ever much the case in the Flint example. There have been historic and systematic policies and practices within the city that often perpetuated the racial segregation of African Americans, which formed the legacy of Flint's water crisis (Michigan Civil Rights Commission, 2017). The implicit biases that existed in policymaking and decision-making regarding the community were not sufficiently critically assessed by those in power to be able to make effective risk judgements that not only keep communities safe but also ensure environmental justice for all.

Expertise and Risk – Should We, as Experts, Be Concerned about Public Apathy in the Face of Growing Risks? Or Should We Be Turning the Lens Back on Ourselves?

> 'Why doesn't anybody believe us anymore?' This question was recently asked by a PhD level biologist, a man who has risen over the years to a position of considerable authority in a federal resource management agency. The specific context was the public insistence that his agency stop

> promoting a 'risky technology', even though the evidence convinced him the risk was low.
>
> (Freudenburg, 1988, p. 242)

This is a passage that could have been written today. William Freudenberg penned this in 1988, and it represents a challenge that remains relevant and intractable for experts in our current society. Is it a problem of a public distrust in expertise, or a lack of introspection – arrogance, even – on the part of experts themselves? Or is it that experts reduce public perceptions as irrational or as a result of information deficit? All done while not turning a self-examining eye on oneself. What underpins the superiority of our own knowledge. How does this frame sway or skew our own risk choices?

Within this book, I discuss potential sources of judgement error that could affect any person, regardless of their 'expert' status. I will further elaborate on how our own probabilistic risk processes are (in Freudenburg's words) obscuring the "person-intensive nature of performing the assessments" (Freudenburg, 1988, p. 242).

After all, risk assessments themselves are performative. They represent an accepted code and ritualised practice that is often fetishised in its supposed ability to produce the objective answer one was seeking. The problem has been assessed, and it is now documented, the performance of quantification has been undertaken, and the answer is clear to all and sundry. The fact that the final justification is inherently a social and political problem and acceptance of such rituals is rarely closely scrutinised, and so often used to justify government spends to pursue solutions along an often pre-determined path. These risk dances are also, not surprisingly, notoriously poor at dealing with uncertainty. Uncertainties might be 'known unknowns', or your Black Swan 'unknown unknowns' (Aven, 2014; Taleb, 2008). Without understanding of your own blind spots, there is little opportunity for turning your head for a different view.

Those of us who do the assessments are just as influenced by biases, cultures and social interactions that will affect our ability to do risk assessments. Some argue that this could result in overconfidence; others question the role of the organisational culture in guiding risk thinking. Furthermore, fiscal and political pressures cannot be easily extricated from such risk debates. Government workers are held to the sway of political posturing, of perceptions regarding suitable spend, and general community value.

With the advent of greater government responsibilities comes an increase in the number of formal quantitative analyses that are created. This also comes with the notion of the changing nature of risks (see Chapter 4 for more on this) as scientific and technological advancements grow ever faster, and risk assessment is often consistently on the backfoot in terms of understanding these new developments. This is because our current risk approaches are heavily based on probability matrices established in the 1950s on nuclear safety. This book begins with an exploration (in the following chapter) of these existing approaches to risk. It is those processes that you likely know well, or have heard about in practice. Academic and research circles have moved away from these dated approaches to risk conceptualisations that extend to the psychological, social and cultural spheres. The chapter argues that risk thought has shifted considerably from these original notions of technical risk approaches (such as those portrayed by international standards ISO31000), and very little is acknowledged within practice regarding other framings of risk. This is potentially dangerous, as it ignores many of the biases or social constructions that can work to sway risk thought, and, in turn, the funds that are allocated to certain projects over others. Intrinsically tied up in risk assessments is a values statement, whether overt or not, of how we believe the world around us should be. As such, these values form an aspect of how we view the importance of human life, livelihoods, autonomy and balance. Importantly, risk often intertwines with moral issues (such as the case of Flint above).

Chapter 3 provides the basis to explore these notions through introducing risk perceptions, a field of study that has blossomed since the 1970s. We continue our risk journey in Chapter 4, in discussing the sociological imperatives of risk. These theories consider us as a society and how we have shifted our thinking in recent years to be focusing on eliminating any and all risk. We move back to understanding water in Chapter 5, outlining the findings of a recent study undertaken on water professionals to capture how they consider risk in their own decision-making approaches. This chapter introduces the topic of risk in water, highlighting current practice and its drawbacks. It also argues that understanding how we perceive risk matters considerably in the way that public-sector decisions are made around water resources.

In the final two chapters, I link the theory to this study to understand the broader notion of how risk drives decision-making in water resource management. I also consider the ways in which risk is prioritised in practice and the implications for funding and innovation. It takes this aspect even further to provide recommendations in shifting risk thought to understanding our inherent biases in decision-making. This is what makes this book so important: it sheds light on our own biases and values so that we can see the decisions we make as practitioners with a fresh set of eyes.

Risk Management for Water Professionals argues that much of the way that risk management is currently undertaken in water is heavily subjective. Feelings, such as those of dread, drive higher risk scores, which result in lower chances of non-business-as-usual projects being pursued. Not surprisingly, feelings of inacceptable risk are more often evoked for new and innovative projects rather than tried and tested approaches. Additionally, when risk is assessed in water decision-making, reputational risk is often overestimated, resulting in a similar reluctance to pursue projects that are new or have never previously been undertaken.

This book is an essential guide for project managers, students and water practitioners to better understand their own decision-making in practice. *Risk Management for Water Professionals* presents a clear pathway to navigate personal biases to make better and more grounded decisions in water management.

References

Aven, T. (2014). *Risk, surprises and black swans: Fundamental ideas and concepts in risk assessment and risk management*. London: Routledge.

Covello, V. T., & Mumpower, J. (1985). Risk analysis and risk management: An historical perspective. *Risk Analysis, 5*(2), 103–120. https://doi.org/10.1111/j.1539-6924.1985.tb00159.x

Craft-Blacksheare, M. G. (2017). Lessons learned from the crisis in flint, Michigan regarding the effects of contaminated water on maternal and child health. *Journal of Obstetric, Gynecologic & Neonatal Nursing, 46*(2), 258–266. https://doi.org/10.1016/j.jogn.2016.10.012

Cuthbertson, C. A., Newkirk, C., Ilardo, J., Loveridge, S., & Skidmore, M. (2016). Angry, scared, and unsure: Mental health consequences of contaminated water in flint, Michigan. *Journal of Urban Health, 93*(6), 899–908. https://doi.org/10.1007/s11524-016-0089-y

Franklin, J. (2015). *The science of conjecture: Evidence and probability before pascal*. Baltimore, MD: Johns Hopkins University Press.

Freudenburg, W. R. (1988). Perceived risk, real risk: Social science and the art of probabilistic risk assessment. *Science, 242*(4875), 44–49. Retrieved from http://www.jstor.org/stable/1702491

Hanna-Attisha, M., LaChance, J., Sadler, R. C., & Champney Schnepp, A. (2015). Elevated blood lead levels in children associated with the flint drinking water crisis: A spatial analysis of risk and public health response. *American Journal of Public Health, 106*(2), 283–290. https://doi.org/10.2105/AJPH.2015.303003

Michigan Civil Rights Commission. (2017). *The flint water crisis: Systemic racism through the lens of flint*. Retrieved from https://www.michigan.gov/-/media/Project/Websites/mdcr/mcrc/reports/2017/flint-crisis-report-edited.pdf?rev=4601519b3af345cfb9d468ae6ece9141

Taleb, N. N. (2008). *The black swan: The impact of the highly improbable*. London: Penguin Books Limited.

The Natural Resources Defense Council. (2018). *Flint water crisis: Everything you need to know*. Retrieved from https://www.nrdc.org/stories/flint-water-crisis-everything-you-need-know#summary

2 Technical Theories of Risk

'Is God Real?' asks Arnobius in a discussion that would underpin one of the first known instances of a risk matrix. In the 4th century AD, he sought to determine the risk associated with devout Christianity, as approach that would later led to 'Pascal's Wager' (see Chapter 5 for more on this). Pascal took Arnobius' proposition and sketched up a simple 4 × 4 matrix with 'God Exists', 'God does not exist' on one axis and 'Be a devout, practicing Christian' and 'Do not be Devout' on the other. Consequences were mapped on the intersecting cells of the matrix. This matrix determined the risk of not gaining admission to heaven on the basis of these two factors. He concluded that the risk of not having a good afterlife was too high, and therefore practising religion was the best risk mitigating strategy. Although our risk processes have changed significantly since that time, the basic underlying premise of the original still remains. This chapter will run through the basic principles of the way that risk is currently measured and quantified. It is driven by what is known as the technical approach to risk and adopts a positivist attitude – viewing risk assessments as an objective exercise undertaken by experts (see Figure 2.1 for where it sits on the conceptualisations scale). The risk assessor first identifies the risks and benefits associated with a project, often gathering data, followed by an objectified assessment based on certain factors devised by experts. Many government organisations and regulatory bodies, such as the Environmental Protection Agency in Victoria, Australia, utilise this form of risk approach to inform the assessment of two inputs: risk consequence and risk probability (Environmental Protection Agency Victoria, 2004) (Kosovac, Davidson & Malano, 2019). This forms a key decision-making tool of a business case and importantly, whether to proceed with a particular project or not.

Starr (1969) was a major proponent of this type of objectified risk analysis. He promoted the idea that risk assessments can be used to form an ideal scenario wherein there is a calculated risk to benefit. This type of assessment relies on legitimate statistical and/or fiscal data being readily available, thus creating an arguably quantified approach to assessing a risk within a project. The underlying assumption is that should a 'rational' person assess the risk, it results in one objective risk score regardless of the assessor. Thus, risk is

DOI: 10.4324/9781003432647-2

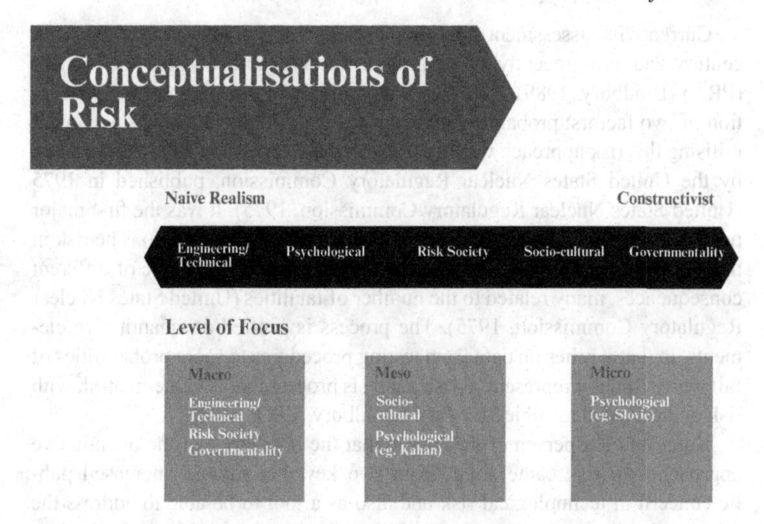

Figure 2.1 Conceptualisations of risk theories and their level of focus.

treated as an objective reality that exists (naive realism), not one that is socially constructed (Figure 2.1).

Probabilistic Risk Assessment and its Discontents

A form of technical risk assessment is the Cost-Benefit Risk Analysis (CBA), which is a central tenet of economic decision-making. The CBA embraces consideration of possible costs versus benefits when assessing future actions, through the prism of economic rationalisation. Heavy reliance is placed on cost assumptions to quantitatively measure pros and cons through a fiscal lens, which may include valuing social benefits, to determine the long-term viability or investment opportunity of a particular option. It often considers depreciation and time-sensitive costings to provide a result in real monetary terms. The focus, therefore, is predominantly on quantifying risk in fiscal lexis. Peter Self's (1975) seminal piece on 'Econocracy' refutes the nature of cost-based risk analyses, and in particular cost-benefit analyses. He argues that there is an inherent falsehood to these approaches that rely on potentially dangerous assumptions and that they should not act as the only guide to decision-making. For Self, cost-benefit analyses and their obsession with costing elements such as welfare and social benefit are flawed and biased, presenting a problematic quantification of risk that obscures value-judgements amongst the figures . In an attempt to generate a more objective and rational approach to decision-making, the method instead creates an assessment that is wholly irrational due to its attempt to quantify elements that cannot necessarily be effectively quantified, thus carrying an inherent risk in its use.

Current risk assessment measurement approaches date back to the 15th century and more recently, its iteration as the Probabilistic Risk Assessment (PRA) (Bradbury, 1989). The PRA was traditionally made up of a combination of two factors: probability and consequence. A key influential document utilising this risk approach was the 'Reactor Nuclear Safety' study undertaken by the United States Nuclear Regulatory Commission, published in 1975 (United States Nuclear Regulatory Commission, 1975). It was the first major project to incorporate the PRA into its planning (Bradbury, 1989). The risk in the report was given by the probability and magnitude of a range of different consequences, many related to the number of fatalities (United States Nuclear Regulatory Commission, 1975). The process is centred on quantitative elements, in that it relies on data from testing procedures to score probabilities of failure. As such, it represents a risk analysis process that is value-neutral, with risk being treated as 'objective fact' (Bradbury, 1989, p382).

Kates and Kasperson (1983) argue that the popularity of the quantitative approaches to risk came about from two key motivations: increased public concern in technological risk and also as a tool to be able to address the ever-challenging area of 'the acceptability of a risk'. I would also add another: responsibility. By having a clear decision-making process, auditors of public accounts can rest happy knowing that there is a system followed by public decision-makers to allocate risk (and money). Technological changes were rife in the 20th century and ensuing public anxieties then required some level of control.

Unsurprisingly, the PRA, (much like similar technical risk approaches) has been criticised in the academic literature for their rigid management of risk which ignores the nature of human beings as non-rational beings (Beck, 1992; Self, 1975). Additionally, it ignores the value-laden decision-making involved in risk analyses. Each decision has an element of the risk assessor's or the organisational values attached to it, as shown when considering organisational risk consequence frameworks. For example, one organisation might see 1,000 people affected by loss of water as being of high risk, yet another may only consider this a medium-level risk. These ratings cannot be said to be objective measures, as they were formulated on an underlying value judgement. Why are 1,000 people considered high risk, why not 500 or even 1? These types of risk ratings can be argued to be somewhat arbitrary in their assessment grading, and therefore the approach altogether cannot be said to be 'objective'.

Risk Matrices as a Risk Assessment Tool

Starting with Arnobius, with significant improvements by Pascal and his probability theorems, the most widely used tool in the technical risk space is the risk matrix (Cox, 2008). This matrix generally considers two key inputs: the consequence of a particular risk scenario and its likelihood of occurring. The consequence focuses on capturing the outcome of an action, and this is scored based on prescribed factors such as threat of injury, fatality, reputation and financial. The likelihood is scored based on the probability of this scenario occurring. These probabilities are also pre-defined and often given scores from 1 to 5 based on likelihood (with 5 as the almost certain), although the scoring systems may change with different risk processes. With the aid of the risk matrix, which is normally devised by the individual organisation itself, consequence and likelihood scores are used to determine the final 'objective' risk score. This approach has been highlighted in documents such as the Australian Standards on Risk Assessments ISO31000, and therefore followed closely by many government departments such as the Environmental Protection Agency and the Department of Health (Australian Government; Environmental Protection Agency Victoria, 2004; Victorian Managed Insurance Authority, 2016).

These risk matrices are still utilised widely, despite the vast criticism of them in academic literature (Cox, 2008; Aven, 2017; Ball & Watt, 2013; Hubbard & Evans, 2010; Pate-Cornell, 2002). These criticisms hinge on a range of factors beginning with the conflation of uncertainty and probability when using risk matrices (Aven, 2017), extending to inconsistencies across assessors in their levels of mathematical probability understanding (Flage, Aven, Zio, & Baraldi, 2014).

A key element of the staying-power of risk matrices exists in its simplicity. As a tool, it can be rolled out across an organisation with diverse functions, while also allowing for a seemingly rational approach of quantifying risk. As such, is a tempting template to use. However, behind these risk measurement tools exist some degree of subjectivity. Decision-making will often carry unexpected or unpredictable outcomes, and risk matrices may not be the most ideal way of measuring the risk. As discussed earlier, the practice of reducing two dimensions (consequence and likelihood) into one dimension is rife across the public sector globally. This carries its own issues in the assumptions that a likelihood score of 3 will be assessed as of equal weight as a consequence of the same value, despite measuring vastly different properties. There are many drawbacks to this approach, namely, that the assessor may not accurately predict the outcome, together with the lack of knowledge to support their assessment of the probabilities of likelihood (Aven, 2017). As risk assessments rely heavily on the score, or rating, received from risk matrices, this inaccuracy in predicting outcomes and likelihood renders the processes problematic and thus can lead to poor decision-making. This is especially pertinent when considering that public sector decision-making is often made on behalf of the community, and could then affect large numbers of people. There are alternatives to risk matrices (see, for example, Ruan, Yin, & Frangopol, 2015; Van Der Sluijs et al., 2005); however, the existing matrix approach still pervades industry practice. Reflecting that uncertainty is a problem, how do we go about measuring risk in increasingly uncertain times?

Decision-Making under Deep Uncertainty

In the last 20 years, there has been a subfield of risk management that has been flourishing: Decision-Making under Deep Uncertainties (DMDU). This area of study specifically targets events that are incredibly difficult to reasonably predict, and often results in risk disagreements across experts (Lempert, 2019). This field has its basis in the field of environmental and infrastructure planning, as both of these reflect areas where there are challenges related to the dependence path of events and the high uncertainty related to their development. Anticipating and planning for future systems, especially in the face of global emergency events such as the Climate Crisis, is increasingly becoming difficult, resulting in large spends on infrastructure projects that may not be resilient to future unexpected shocks. The DMDU research field instead bases its methods on a 'monitor and adapt' scenario, rather than a 'predict and act' one (Stanton & Roelich, 2021). Proponents argue that in this way it works to embrace uncertainty.

Jan Kwakkel, a key proponent in the DMDU field (along with their colleagues), first describes uncertainty as being '*limited knowledge* about future, past, or current events' and then refutes that very definition by stating that "Uncertainty is not simply the absence of knowledge… [which] can be of three sorts: inexactness, unreliability, and border with ignorance" (Walker, Lempert, & Kwakkel, 2013, p. 293). If the probability is not a 0 (will definitely not happen) or a 1 (certain to occur) it can be considered uncertain. Technical approaches to risk (as previously described in this chapter) have pre-supposed a naïve positivist stance on uncertainty, which relies on constant information seeking to determine the 'true' value of risk (Lupton, 2013; McDaniel & Driebe, 2005). However, all risk, much like all uncertainty,[1] exists in the future; it follows on that nothing can genuinely be certain, a supposition that Walker et al. (2013) also agree with. Uncertainty is a condition, a fallible human state that has its origins in the Middle English 'uncerteynte', c. 1300, meaning not fully confident. This lack of confidence does not necessarily presuppose a lack of information, or a lack of knowledge, but is a human predisposition, a human state that may not be borne off with figures.

The uncertainty that DMDU deals with is termed as Level 4 and 5 uncertainties, i.e. those that are 'known unknowns' and those where analysts do not know or can agree on the outcome (Walker et al., 2013). It veers away from naïve realist 'information seeking' but still remains firmly in the epistemological and ontological surrounds of this space (see Figure 2.1). Although recognising the fallibility of humans *to some extent*, DMDU's main debates are positioned in risk definition, risk measurement, accuracy of data and expert consensus. The approach does not question the predictive models themselves, but the ways that they are used (Walker et al., 2013). There are many tools that a DMDU proponent may use to address risk for Level 4 and 5 systems. This most often includes the use of scenario planning, which involves determining a range of likely outcomes (or scenarios) for assessment. Similarly, adaptive planning to consistently address and reanalyse situations on the basis of new information is also used to deal with these levels of uncertainties. This approach plans for a constant review of decisions at regular intervals, rather than undertaken on an ad hoc basis (Walker et al., 2013).

Robust Decision-Making (RDM) is another tool in DMDU's arsenal. Focusing on questioning assumptions, the process aims to discount certain pathways over others on the basis of 'what if' statements (Lempert, 2019). An example that was used to demonstrate this point is a 25-year plan for water resources in Southern California. Initially created in 2005, the plan did not consider future uncertainties that came to the fore regarding water, namely, climate change. Taking an RDM approach meant determining the various assumptions used in initial plans, and developing scenarios wherein these assumptions are tested. The process found that a decrease in precipitation, groundwater and failure to produce sufficient recycled water would result in an overall plan failure. This led analysts to reconsider this failure pathway to reduce the amount of vulnerability in the system. Although this does take a distinct sidestep from other techno-scientific approaches, there are a number of factors that expose it to a full gamut of influences. 'How are risks constructed as social facts?' asks Lupton (2013), a question not adequately addressed in any of these processes, a question that arguably produces the largest amount of uncertainty. Although there is acknowledgment that humans are imperfect decision-makers, there is little in the models to understand or effectively account for this, leading to a glaring omission (and large vulnerability in its approach). This is an element that is also seen in the resilience literature.

Resilience Thinking: A False Sense of Readiness

As DMDU theories were increasing in use, similarly did we see a burgeoning in the popularity of resilience thinking both in academic circles and in practice. Resilience's greatest benefit is similarly its greatest detractor: its ambiguity. Resilience is often mischaracterised as wide-ranging definitions abound that lead to inconsistencies in approach (Gleeson, 2013; Kosovac & McLeod Logan, 2021). This is a problem in ensuring effective planning, as two people sitting across the table from one another may very well have differing conceptualisations of the word. Similarly, the increasing usage of the term in parlance means that it is also often conflated with psychological resilience. Many scholars have proposed their own definitions of the term (see, for example, Fastenrath & Coenen, 2021; Logan, Aven, Guikema, & Flage, 2021), but there is yet to be an overarching one that dominates the planning and risk spheres.

So, one problem is its definition, and the other problem is in its use. As I have argued elsewhere with my colleague, Dr Tom Logan (Kosovac & Logan,

2021), systems are often described as being resilient, or given a goal of future resilience (see for example, The Office for the Minister of Foreign Affairs (2019) or the 100 Resilient Cities Programme). The implication is that being resilient is achievable and desirable. If the perceived goal of being 'resilient' is reached by decision-makers, could this imply a false sense of security? Most concerningly, I argue, this could manifest if individuals or a society believes they are *resilient* (to all hazards and uncertainty), as a result of being resilient to a single (type of) hazard.

General resilience perpetuates the idea of resilience in its most comprehensive form (Carpenter et al., 2012). Advocates contend that general resilience signifies the ability to withstand various types of shocks, even those that are novel and unexpected (Carpenter et al., 2012). They assert that general resilience is attained by strengthening various essential factors they have pinpointed, such as leadership, diversity and modularity. However, in social systems, the idea of general resilience is potentially flawed, in a manner consistent with the issues surrounding the 'all-hazards approach' to hazards. The all-hazards approach is "*an integrated approach to emergency preparedness planning that focuses on capacities and capabilities that are critical to preparedness for a full spectrum of emergencies or disasters*" (CMS – Centers for Medicare and Medicaid Services, 2019). It is criticised predominantly as the differences required to respond to different hazards can often be quite vast and the commonalities may not be simple (Bodas, Kirsch, & Peleg, 2020). And yet, these commonalities are the basis of general resilience. The issue with both the all-hazards approach to risk and the notion of general resilience is the potential for a false sense of readiness that results in suboptimal preparation. This false sense of readiness manifests in a feeling of security that may be endangering the safety of many, especially when considering large-scale disaster risk.

How resilient is resilient enough? The parallels between resilience and safety are striking and perhaps characterising resilience as the analogy to community or societal safety is what is needed. This would require factoring in aspects of recovery, robustness and long-term adaptability. However, to do so would enable the concept to be more readily understood and recognised for its inherent subjectivity. If we accept the definition that *a system is judged resilient if the risk of not achieving the desired functionality is sufficiently low* then that judgement is based on the knowledge, values and aversions of the individual or group making that decision. This inherent subjectivity of resilience requires that care is taken around how the concept is communicated to avoid creating false security or misperception. This is especially critical in community resilience settings where the decision-makers are often not experts in risk or hazard analysis. This subjectivity also raises ethical and social justice questions around what is valued by the community, over what time frame is risk considered and who is making these decisions.

Rational Actor Theory and Technical Risk

Both the technical and economic risk assessments rely on the assumption that all decision-makers and actors are rational. Through utilising Rational Actor Theory when predicting the consequences and actions of others, technical risk proponents argue this ensures a positivist and objectified approach to risk management. Rational Actor Theory is based on actors taking action based on predictable, consistent goal setting while also maximising expected utility. Von Neumann and Morgenstern's (2007) expected utility theorem posits that a decision-maker, when faced with potentially risky scenarios, will have a preference for actions that maximise expected utility. Any actor that does not fit this decision-making paradigm is said to be 'irrational'. The rational actor is diametrically opposed to the irrational actor, who is deemed to be spontaneous, random and impulsive. The rational actor is assumed to be acting in self-interest. Many, such as Rothschild (1946), argue that the assumption of rationality is essential in modern economic theory; however, Arrow posits that this is an entirely unnecessary assumption to make in decision theory as "we could just build up our economic theory on other assumptions as to the structure of choice functions if the facts seem to call for it" (Arrow, 2012, p. 21).

Due to limits of rationality in individuals, it cannot be assumed that all are rational actors in their decision-making. As purely rational actors are fictional, it remains necessary to leave space for an element of subjectivity in the conception of risk. Slovic (1993) argues that there is simply no such concept as a 'real risk' or 'objective risk'. He argues that in assuming that the probability combined with the consequence of risk is objective, and that all assessors are rational in their approach, it ignores the very nature of a 'perceived risk'. This assertion hinges primarily on the claim that the way in which a layperson would analyse a risk is likely to differ from the approach taken by an expert in that field, due to the tools one has on hand.

These tools of analysis can be drawn from factors such as education, culture and lived experience in order to determine one's relationship to a risk. The sense of dread associated with a risk, as discussed by Slovic (1993), accounts for a strong emotive response to a risk in decision-making. The higher the factor of 'dread risk', the larger the perceived risk as analysed by the public. For instance, Slovic argues that phenomenon which he terms 'dread risks', which often loom large in the public consciousness due to previous catastrophes (a paradigm example of which would be nuclear power), is likely to be perceived as posing a high level of risk by lay people, even in cases in which this assessment is not reflective of expert opinion (see Chapter 3 for more on this). This may not fall in line with experts' assessments of risks when based on a technical risk analysis. This is due to the fact that current risk assessments are positioned to base their perceptions of risk on potential mortalities or level of harm per capita. The public frequently reject models of risk devised by scientists using these methods, as they are unlikely to adhere to the value-based perceptions presented by an individual. This is not to state that a particular viewpoint, whether it be the layperson or the expert, is necessarily better than the other, but rather to highlight that there are elements that both groups may not consider that the other group will place importance upon (see Chapter 6 for more on expertise, risk and community trust). These risk analyses also fail to effectively take account of social and political elements of risk, which can be difficult to measure quantitatively. Therefore, the argument that technical perspectives on risk are risky themselves is apt when considering the enormous effect that social impacts may have on a project/process.

The following chapters will diverge from the quantified risk approaches we currently see in practice. While technical theories of risk see risk as being 'real' and able to be 'known', other theories digress from these notions. Psychological theories of risk do tend to see risks as 'real', but admit that we can never truly know them (even though we may get close at times). Many of the sociological theories place risk as a partially or fully constructed reality, which we can never truly know as it is not there to be 'found'. It is created for and by us. These theories can help us begin to sort through the varied notions of risk, and inform our own risk understandings in future.

Note

1 A note on risk vs uncertainty: Knight, 1921 denotes risk as being a calculable and controllable aspect of the unknowable. Uncertainty, he argues, is the remainder, not being able to be calculated and not being able to be controlled.

References

Arrow, K. J. (2012). *Social choice and individual values* (3rd ed.). New Haven, CT: Yale University Press.

Australian Government, D. o. I., Innovation and Science. Quantification of consequence/likelihood matrices. Retrieved from https://archive.industry.gov.au/resource/Programs/LPSD/Risk-management/Appendices/Appendix-1-Risk-analysis/Pages/Quantification-of-consequencelikelihood-matrices.aspx

Aven, T. (2017). Improving risk characterisations in practical situations by highlighting knowledge aspects, with applications to risk matrices. *Reliability Engineering & System Safety, 167*, 42–48. https://doi.org/10.1016/j.ress.2017.05.006

Ball, D. J., & Watt, J. (2013). Further thoughts on the utility of risk matrices. *Risk Analysis, 33*(11), 2068–2078. https://doi.org/10.1111/risa.12057

Beck, U. (1992). *Risk society: Towards a new modernity*. London: Sage Publications.

Bodas, M., Kirsch, T. D., & Peleg, K. (2020). Top hazards approach – Rethinking the appropriateness of the all-hazards approach in disaster risk management. *International Journal of Disaster Risk Reduction, 47*, 101559. https://doi.org/10.1016/j.ijdrr.2020.101559

Bradbury, J. A. (1989). The policy implications of differing concepts of risk. *Science, Technology, & Human Values, 14*(4), 380.

Carpenter, S. R., Arrow, K. J., Barrett, S., Biggs, R., Brock, W. A., Crépin, A.-S., … Zeeuw, A. D. (2012). General resilience to cope with extreme events. *Sustainability, 4*(12), 3248–3259. https://doi.org/10.3390/su4123248

CMS - Centers for Medicare and Medicaid Services. (2019). *Emergency preparedness-updates to appendix Z of the state operations manual.* Retrieved from https://www.cms.gov/Medicare/Provider-Enrollment-and-Certification/SurveyCertificationGenInfo/Downloads/QSO19-06-ALL.pdf.

Cox, A. L. (2008). What's wrong with risk matrices? *Risk Analysis, 28*(2), 497–512. https://doi.org/10.1111/j.1539-6924.2008.01030.x

Environmental Protection Agency Victoria. (2004). *Risk-based assessment of ecosystem protection in ambient waters* (Publication 961). Melbourne. Retrieved from http://www.epa.vic.gov.au/~/media/Publications/961.pdf

Fastenrath, S., & Coenen, L. (2021). Future-proof cities through governance experiments? Insights from the Resilient Melbourne Strategy (RMS). *Regional Studies, 55*(1), 138–149. https://doi.org/10.1080/00343404.2020.1744551

Flage, R., Aven, T., Zio, E., & Baraldi, P. (2014). Concerns, challenges, and directions of development for the issue of representing uncertainty in risk assessment. *Risk Analysis, 34*(7), 1196–1207. https://doi.org/10.1111/risa.12247

Gleeson, B. (2013). *Resilience and its discontents, Research Paper No. 1.* Retrieved from Melbourne: https://sustainable.unimelb.edu.au/__data/assets/pdf_file/0006/2763492/MSSI-ResearchPaper-01_Gleeson_Resilience_2013.pdf

Hubbard, D., & Evans, D. (2010). Problems with scoring methods and ordinal scales in risk assessment. *IBM Journal of Research and Development, 54*(3), 2:1–2:10. https://doi.org/10.1147/jrd.2010.2042914

Kates, R. W., & Kasperson, J. X. (1983). Comparative risk analysis of technological hazards (A review). *Proceedings of the National Academy of Sciences of the United States of America, 80*, 7027–7038.

Kosovac, A., Davidson, B., & Malano, H. (2019). Are We Objective? A Study into the Effectiveness of Risk Measurement in the Water Industry. *Sustainability, 11*, 1279. https://doi.org/10.3390/su11051279

Kosovac, A., & McLeod Logan, T. (2021). Resilience: Lessons to be learned from safety and acceptable risk. *Journal of Safety Science and Resilience, 2*(4), 253–257. https://doi.org/10.1016/j.jnlssr.2021.10.002

Lempert, R. J. (2019). Robust decision making (RDM). In V. A. W. J. Marchau, W. E. Walker, P. J. T. M. Bloemen, & S. W. Popper (Eds.), *Decision making under deep uncertainty: From theory to practice* (pp. 23–51). Cham: Springer International Publishing.

Logan, T. M., Aven, T., Guikema, S., & Flage, R. (2021). The role of time in risk and risk analysis: Implications for resilience, sustainability, and management. *Risk Analysis.* https://doi.org/10.1111/risa.13733

Lupton, D. (2013). *Risk* (2nd ed.). New York and Oxon: Routledge.

McDaniel, R. R., & Driebe, D. J. (2005). *Uncertainty and surprise in complex systems: Questions on working with the unexpected.* New York: Springer.

Pate-Cornell, E. (2002). Risk and uncertainty analysis in government safety decisions. *Risk Analysis, 22*(3), 633–646. https://doi.org/10.1111/0272-4332.00043

Rothschild, K. W. (1946). The meaning of rationality: A note on Professor Lange's article. *Review of Economic Studies, 14*(1), 50.

Ruan, X., Yin, Z., & Frangopol, D. M. (2015). Risk matrix integrating risk attitudes based on utility theory. *Risk Analysis, 35*(8), 1437–1447. https://doi.org/10.1111/risa.12400

Self, P. (1975). *Econocrats and the policy process: The politics and philosophy of cost-benefit analysis*. London: Macmillan.

Slovic, P. (1993). Perceived risk, trust, and democracy. *Risk Analysis: An International Journal*, 675. Retrieved from https://ezp.lib.unimelb.edu.au/login?url=https://search.ebscohost.com/login.aspx?direct=true&db=23h&AN=33672416&site=eds-live&scope=site

Stanton, M. C. B., & Roelich, K. (2021). Decision making under deep uncertainties: A review of the applicability of methods in practice. *Technological Forecasting and Social Change, 171*, 120939. https://doi.org/10.1016/j.techfore.2021.120939

Starr, C. (1969). Social benefit versus technological risk. *Science, 165*, 1232–1238.

The Office for the Minister of Foreign Affairs. (2019). *Stepping up climate resilience in the pacific*. Online: The Australian Government. Retrieved from https://www.foreignminister.gov.au/minister/marise-payne/media-release/stepping-climate-resilience-pacific

United States Nuclear Regulatory Commission. (1975). *Reactor safety study: An assessment of accident risks in U.S. commercial nuclear power plants*. Retrieved from https://www.nrc.gov/docs/ML0706/ML070610293.pdf

Van Der Sluijs, J. P., Craye, M., Funtowicz, S., Kloprogge, P., Ravetz, J., & Risbey, J. (2005). Combining quantitative and qualitative measures of uncertainty in model-based environmental assessment: The NUSAP system. *Risk Analysis, 25*(2), 481–492. https://doi.org/10.1111/j.1539-6924.2005.00604.x

Victorian Managed Insurance Authority. (2016). *The Victorian government risk management framework practice guide*. Retrieved from https://www.vmia.vic.gov.au/risk/risk-tools/risk-management-guide

Von Neumann, J., & Morgenstern, O. (2007). *Theory of games and economic behavior [Electronic resource]*. Princeton, NJ: Princeton University Press.

Walker, W. E., Lempert, R., & Kwakkel, J. (2013). Deep uncertainty. In S. Gass & M. C. Fu (Eds.), *Encyclopedia of operations research and management science* (3rd ed., pp. 395–402). New York: Springer.

3 Psychological Theories of Risk

Socrates (S):	… Isn't it sometimes the case that, when the same wind is blowing, one of us shivers, the other doesn't? And that one of us shivers slightly, the other a lot?
The aetetus (T):	Very much so.
S:	Then are we going to say that on such occasions the wind is cold or not cold in itself? Or are we going to believe Protagoras – that it is cold for the one who shivers, not cold for the one who doesn't?
T:	We should, it seems.
S:	And is that not how it appears to each of them?
T:	Yes.
S:	And 'appears' is the same as 'perceives'?
T:	Yes, it is.
S:	Then appearance and perception are the same thing, both in questions of heat and in all such matters. For the way each person perceives things also looks like being how they are for each person.

(Plato, *Truth*, 152b2–c4, as quoted in Sedley, 2004)

Knowledge is perception. This is the main argument under scrutiny in the above (Plato's Theaetetus). That X exists as true for the perceiver, which cannot be refuted by another, hints at the notion that there is minimal space for non-perceptual knowledge (Sedley, 2004). The fallibility of humans means that one can never truly 'know'. The knowledge surrounding risk, and perception of it, is an area that psychologists have been grappling with since the 1960s. "Danger is real, but risk is socially constructed" states prominent psychological theorist Paul Slovic (1993, p. 659). Similar to the technical risk approach, psychological theorists tend to believe that 'real' risk exists, but acknowledge that we can never truly know it, as our minds will always skew the inputs. So the two risk approaches are ontologically in agreement (is risk real?), yet divergent in their epistemology (can we ever truly know it?) (see Figure 2.1 in Chapter 2). Psychological proponents state that our ability to

DOI: 10.4324/9781003432647-3

see the 'true' risk is always shifted by our own psychological makeup. Lying in contrast to the technical view discussed in Chapter 2, the unit of measure for the psychological view zeros in on the individual and how they form judgements based on their underlying beliefs and experiences. Subsequently, risk perception is closely tied to an individual's view, thus linking risk choices to their own values and psychology (Fischhoff, Slovic, Lichtenstein, Read, & Combs, 1978). These experiences create a personal prism through which the individual assesses risks.

Emerging in the 1960s and 1970s, psychological theories of risk were a direct response to the *raison d'entre* of risk assessments at that time: probability-based theories and matrices. Scholars such as Paul Slovic and Baruch Fischhoff were critical of these risk appraisal methods, calling upon discrepancies between what was being measured, and how people viewed the risk themselves. They started asking questions about why some people were afraid of nuclear power rather than driving, despite the statistics around driving killing countless more people. Their work, along with many other psychological and decision scientists, started exploring this disconnect. In particular, they explored decision-making biases in people, particularly when it comes to processing uncertainty. These exist as a result of personal and contextual variables that steer an individual to assign risk importance for certain factors over others. In their seminal paper on this topic, Daniel Kahneman and Amos Tversky (1974) use 'heuristics', known as decision-making shortcuts, to explain decision-making in times of uncertainty, which they argue is most of the time, as information received is ordinarily incomplete. There are a number of these heuristics that impact on the way that an individual might perceive a risk. They argue that we use two systems (1 or 2, or both) to make day-to-day decisions and these heuristics form part of the System 1 approaches to psychological responses. System 1 is fast, immediate and relies on intuition and automatic responses. System 2 instead is slow, deliberate and requires conscious effort. Tversky and Kahneman (1974) write that people often rely on a combination of System 1 and System 2 when perceiving risks. The problem with System 1, however, is that it is based on several biases. A System 1 appraisal relies heavily on making a decision quickly, so this requires our minds to pull data from its closest source: ourselves. There are a number of different heuristics that psychologically influence risk decisions specifically, including the availability heuristic, optimism, representativeness and confirmation bias. These are explained in more detail below.

Availability Heuristic

Think about a situation where you hear that three close contacts around you have recently had their houses broken into. When asked about the risk of your own house being broken into, you might recall these events relayed by your friends, and understandably respond that you feel that the risk of getting your house broken into has increased. This is without searching for any available evidence regarding the instances and trends of house break-ins. What has happened here is that your mind has quickly pulled up the memory of your contacts' distressing venting of their experience. This is the availability heuristic in action. The availability heuristic refers to a phenomenon that occurs when people are able to judge an event as more likely when it is easier to recall, despite their exposure to any data that may contradict their perceptions (Fischhoff, 1975; Taylor, 1982).

> A person is said to employ the availability heuristic whenever [they estimate] frequency or probability by the ease with which instances or associations come to mind.
>
> (Tversky & Kahneman, 1974, p. 208)

But how does this relate to our risk assessments? This decision-making shortcut helps explain the mental calculation of likelihood by risk assessors, as the ease by which an individual can bring up examples of a particular hazard will determine the likelihood score they assign to that hazard (Taylor, 1982). The easier it comes to mind, the higher the perception of likelihood of occurrence. The interesting aspect of the availability heuristic is that it can be swayed or affected by a myriad of inputs. This could be via social media (seeing a greater prevalence of a certain type of hazard occurring), through news reports, conversations with friends or even your own experience. The impact will vary according to affective responses. The sense of a feeling you get about the experience, especially if it evokes a feeling of dread, will also render the risk seemingly more likely to occur.

The elevation of particular risks as a result of an individual's awareness or experience of them can also influence the resources the individual allocates to ameliorating that risk, compared to other risks they have no personal affiliation with. This scenario creates a disparity in the homogeneity of behaviours towards a particular risk, which can in turn promote division and debate, leaving the definition of the risk itself open to interpretation.

Optimism – Is It Always a Good Thing?

Faced with a flooded road, a person decides to drive through it. They have personally never been swept away by floods, and they believe that despite the warnings, they are not at risk. Why is this? A large part of it is associated with believing that you are less likely to experience a negative event compared to others around you. That, in general, good things will happen to you. This bias can lead to an under-assessment of risk. It may prompt this person to drive through dangerous floodwaters, waters that can pick up a car at just 30cm (1 ft) depth. This is called the optimism bias. Not only is the optimism bias a culprit in our underappraisal of risk, but it also has a more marked effect on our assessment of negative events, leading us to underestimate these events more than positive ones (Langer, 1982).

Another factor that changes our perception of a risk is the representativeness bias. The representativeness bias uses our preconceived notions of what we anticipate, to inform our risk perceptions, i.e. "does this thing look or act like I would expect?" This sways our risk perceptions, as it is often not based on facts or evidence, but rather our own experience, or our own idea of expectations and stereotypes surrounding the event. For example, consider a community that has experienced a series of extreme weather events such as floods over a short period of time. After these events, individuals may develop a representativeness bias by assuming that these extreme weather occurrences are the 'new normal' for the region. They 'expect' more floods in future.

As a result, this is likely to affect individual's risk decision-making and preparedness based on these experiences. For example, they may invest in flood-resistant homes or relocate to what they believe are safer areas, without considering the broader historical climate data or consulting with environmental experts. They may also assume that because they have recently been faced with extreme weather events, this pattern will persist indefinitely. However, this assumption overlooks the complexities of climate science and the variability in weather patterns over time. It may lead individuals to overestimate immediate risks or underestimate the potential long-term impacts of climate change in different regions. This representativeness bias can influence decisions related to environmental risk management without considering a more comprehensive and nuanced understanding of environmental factors.

This is also closely linked to another heuristic: confirmation bias. This is where we seek information that confirms our prior beliefs of the risk (think about the 'expectations' of the representativeness bias and how it links to information-seeking). For example, if you were looking for the risks of fluoride in water, with a preconceived notion that fluoride in drinking water was wholly unsafe, then confirmation bias would lead you to use search terms such as 'unsafe' and 'fluoride', terms which are more likely to bring up anti-fluoride sources. Similarly, discounting information that is not fitting to one's preconceived notion is another way for this bias to take effect on our risk perceptions (Tversky & Kahneman, 1974).

You can probably think of a number of ways where the above heuristics have played out in your own life. It is rare to find anyone that is wholly immune to such effects. Considering these heuristics, Paul Slovic, Baruch Fischhoff and Sarah Lichtenstein, among others, have determined some of the key characteristics by which to measure perceived risk in individuals. As previously discussed, they sought to determine how one can 'see' a nuclear risk as being more dangerous than a driving risk. They also asked themselves what 'safe enough' means for a community, particularly when considering new technologies (Fischhoff et al., 1978). It is, at its core, a balancing of benefits vs the risks. This is a common question asked of water policymakers and practitioners when implementing new systems, or even the upgrading of current systems: how safe is safe enough? While Slovic, Fischhoff and Lichtenstein (1981) recognise that much of this work has stemmed from the economic realm (e.g. cost-benefit analysis and 'revealed preference' models), they turn it on its head to recognise the psychological biases that seek to mould these preferences. These are not biases that can be easily 'revealed' through these economic models as they were heavily based on fatalities up against expected increase in capital (Starr, 1969). You can start to see the worrying elements of a measurement such as this. Psychological theorists thus sought to create ways to bring in heuristics, and psychometric measurements into understanding risk acceptance.

Risk as Feelings

There are eight key characteristics within the psychometric paradigm (Fischhoff et al., 1978) that have been linked to affecting perceived risk:

- Voluntariness and control
- Immediacy of death
- Knowledge of risk
- Newness of risk
- Chronic or one at a time
- Dread
- Fatal

Emotions that indicate a positive or negative feeling are key factors in effective decision-making in risk assessments. The psychological theory of risk takes into account the role of emotions in decision-making, an area that has been long scrutinised by psychologists and ignored by other fields (only recently considered by sociologists analysing risk) (Zinn, 2008). This is heavily reflected in scholarly areas such as the 'risk as feelings' literature (see for example, Loewenstein, Weber, Hsee, & Welch, 2001; Sjöberg, 2000; Slovic, Finucane, Peters, & MacGregor, 2004). The 'risk as feelings' approach gives preferential treatment to risk consequences rather than the likelihood of the risk. An example of this is a terrorist attack. When called to mind, sentiment towards the risk of terrorism tend to provoke feelings related to the outcome of the attack, rather than its chance of occurring (Sunstein, 2003). Subsequently, feelings tend to add significant weighting to the consequences rather than the probabilities of a risk. This does not presuppose irrationality, or a decline in decision-making ability. In fact, approaches that consider past emotional experiences would appear to be imperative in making reasonable and sound decisions (Baumeister, Dewall, & Zhang, 2007). If a strong emotional connection to risk exists, it can cloud an individual's capacity to identify the probabilities of certain scenarios, focusing instead on the possible outcomes and the dread factor associated with them.

Breaking Down the Psychometric Factors

Starting with voluntariness of risk, the sense of being able to opt in or opt out of a risk often presents itself as a key distinction in how high or low a risk is perceived. People tend to see an action as being riskier if exposure is not voluntary. Starr's (1969) work first highlighted the effect of voluntariness upon a risk, stating that the public is 1,000 times more likely to accept a voluntary risk rather than one that is seen as involuntary. In other words, people are generally more likely to take risks if it is their own choice and less likely to accept even a lower level of risk if they feel that it is forced on them (Langer, 1982). Closely tied to the 'voluntariness' of a risk, the risk attribute of control, or the illusion of it, also has an impact on personal risk perception (Langer, 1982). This goes back to Slovic's question raised earlier about why people were so afraid of nuclear power, or flying in an aeroplane, over and above driving their car, despite the fact that driving had a much higher incidence of fatalities. If someone feels a greater control over the danger, then (generally) the lower the perceived risk. Similarly, the control over the risk as well as the voluntariness of exposure in tandem creates a compelling case for discerning the difference between the perceptions of driving risk vs nuclear risk.

Similarly, Langer (1982) assesses the risk perception of those throwing a die, noting that the risk acceptance is higher when the person throws the die themselves rather than someone else undertaking this action. While the probability does not change, the sense of control does, thus guiding risk acceptability behaviour.

Most of us have some level of control over our diets, and as such we often see eating unhealthy foods as a lower risk than many other more overt hazards. Although heart disease is a leading cause of death, the high level of perceived controllability and exposure, combined with its slow impact, leads many to see it as a low risk item. As such, a person's perception of an event changes depending on whether ill health is expected to occur immediately or at some later date. Another example could be the ongoing effect of pollution. Although day to day it may not have a noticeable impact, it could shorten lifespans over a longer period of time. Previous research has also shown that people are more likely to have a higher risk perception for a hazard that kills many people at once (such as an attack) versus one that kills the same number of people over many years (such as diabetes) (Bodemer, Ruggeri, & Galesic, 2013).

Somewhat unsurprisingly, fatalities increase the perceived likelihood of a risk (Slovic, Fischhoff, & Lichtenstein, 1979). If perceived risk rises, its seeming benefits often drop. Tversky and Kahneman (1974) demonstrated that individuals are risk averse if the potential losses are high and risk prone if the stakes for gains are high. Therefore, risk assessments are closely correlated to a level of loss aversion (Fischhoff et al., 1978). Fischhoff et al. (1978) highlight the difference between lay and expert risk perceptions, in particular that expert risk perceptions were more likely to be based on the number of fatalities, whereas the public may be more likely to consider consequences based on vivid imagery. Consequently, where only experts in water are being assessed, based on this literature, one would expect that the likelihood of fatality would play a role in the way that risks are perceived, and in turn on the risk scores those assessors calculate.

Our knowledge (both individually and known by science) will change how we see a risk. Lower knowledge of a hazard is often perceived as riskier versus those we know little about (Fischhoff et al., 1978). As such, the amount of knowledge one has influences the acceptability and decision-making around risks (Kahan et al., 2012; Slovic, Peters, Grana, Berger, & Dieck, 2010). Slovic et al. (2010) explored this in the context of prescription drugs and how the public perceives these related risks. They found that the way that the drugs were perceived was closely tied to their own perception of the knowledge they had about the drugs. (Slovic et al., 2010). This characteristic is driven by a sense of uncertainty and the lack of ability to draw on experience to form a reliable psychological risk assessment of the event occurring (Skjong & Wentworth, 2001).

A feeling of dread is addressed considerably in psychological risk perception theories. In fact, a strong emotional feeling of dread has been shown to be one of the largest factors in propelling an individual to act upon a risk (Baumeister et al., 2007). This feeling is typically associated with personal impacts, impacts on future generations and catastrophic widespread risks (Baumeister et al., 2007). The dread response is closely linked to an imagined outcome of an event or action prompting strong emotional affect. These events often have high consequence and low likelihood perceived risks. Gigerenzer (2004) explores the psychology of dread risks in his research on risk perceptions around terrorist attacks concluding that people would rather drive than fly to a destination in order to avoid an attack similar to that which occurred on September 11, 2001 (the dread risk) (Gigerenzer, 2004). There would be many more lives lost if people had avoided the risk of flying by driving to their destination. Despite the incidence of fatalities on the road being higher than if one was to fly, the inherent psychological effect of the dread risk renders many to still see driving as safer than flying.

A risk that also affects people in an unfair way is likely to be judged higher risk than one that is considered to be just (Keller & Sarin, 1988). In this context, fairness is determined by whether the consequence of the risk affects everyone equally or whether it detrimentally affects a smaller group, therefore deeming the risk 'unfair'. This is closely aligned with whether there is 'choice' involved in the exposure to risk.

Finally, although all of these psychometric factors have been explored within various research studies, there is a methodological issue in that no standardised measurement exists across the field. This has led to difficulty in comparing perceived risk and groups across bodies of research. Wilson,

Zwickle and Walpole (2019) reflect upon this challenge, and seek to understand which of the methods are most effective in capturing personal risk imaginings. They conducted a study using a range of approaches and had concluded that there were three clear measurements that had the strongest impact on risk perceptions: affect, consequence and probability. In particular, the two that linked to personal experience (affect and consequence) proved to have the strongest effects on how one perceives a risk. Several studies only calculated probability to measure risk perceptions, an approach that is "much less critical to individual's subjective assessments of risk" (Wilson et al., 2019, p. 787). As a result, recognising that people will focus predominantly on risk outcomes helps to not only understand perceptions but also guide risk communications.

Risk Perceptions, and How these Differ between Laypeople and Experts

'[The Australian Government] is sick of experts' declared former Australian Department of Human Services Secretary, Renee Leon. She notes that "we have seen an attack on expertise in the last decade where to be an expert was almost to be reviled for being part of an elite of people" (Rollins, 2020). If expert advice did not align with the views of the Government, Leon reportedly maintains, officials preferred to instead rely on "their more favoured decision-making input, which is anecdote" (Rollins, 2020). Disturbingly, this rhetoric is not uncommon, and represents statements mirrored by politicians such as the former US President Donald Trump and prominent UK Minister Michael Gove (Gadarian, Goodman, & Pepinsky, 2020; Riechmann & Madhani, 2020). It is not unusual to see these anti-science narratives framed in a rhetoric of elitism that serves to further cement a harsher distinction between scientific expertise and lay audiences, thus rendering mutual understanding of risk increasingly difficult to coalesce.

Extensive research has been undertaken on the differences in risk perceptions between laypeople and experts. Predominantly undertaken in the United States, this field of study expanded in the 1980s and continues to provide insights into understanding public perception, particularly in an environment in which participatory democracy is encouraged. When comparing public risk perceptions to that of experts, Slovic et al. (1983) showed that laypeople surveyed generally ranked nuclear energy high in risk because of the sense of dread associated with the effects of a negative radioactive fallout. In comparison, laypeople ranked the risk of driving a car very low, despite figures showing the chance of being involved in a car accident is far higher than being caught up in a nuclear one (Slovic, Fischhoff, & Lichtenstein, 1985). Experts instead ranked driving a car as significantly higher risk than that surrounding nuclear power, which the authors suggest is based on an objective understanding of the safety of nuclear energy. Thus, they concluded that the dread or familiarity factor for experts is dissimilar to the layperson.

This difference in risk perceptions evident between laypeople and experts becomes a challenge when the question of public policy, or deciding on public spending, enters the discussion. Prior to the 1960s, expert assessments were generally accepted by the public, with far less scrutiny than they are today (Nichols, 2017). As such, public consultations and public opinion polling did not carry the relevance it currently does which now forms an integral part of project planning. Schon (1995) argues that the loss of faith in professional judgement occurred between 1963 and 1981, much due to the unintended side effects of new technologies and the conflicting recommendations within the scientific field. The consistent debunking of scientific advice and theories also proved to be a large factor in the increasing distrust of professional experts (Nichols, 2017; Schon, 1995). Slovic (1993) also contends that this increase in distrust has flowed on to include public distrust towards risk assessments conducted by these experts.

To show this, a study by Schlosberg (2017) in Melbourne, Australia has shown a clear distinction in how climate change is discussed amongst government authorities and within the public sphere. In considering climate change adaptation, local government bodies used words like risk, water, control, event and management. In contrast, they found that community groups used language that focused more on the impacts of basic needs, such as 'food', 'community', 'people', 'energy', 'water' and 'local' (Schlosberg et al., 2017). The authors highlight the discordant nature of risk discourses between laypeople and experts, and posit that this may be due to lack of public engagement, while others (Guy, Kashima, Walker, & O'Neill, 2014) theorise that it is a knowledge-based distinction.

Reported instances of public dissent and hostility towards expertise may exacerbate issues already at crisis point (Calisher et al., 2020). A denial of facts and of evidence-based decision-making is troubling at the least, and catastrophic at worst. The seeds of doubt towards expertise that were planted by influential public figures in the COVID-19 pandemic cost thousands of lives due to widespread public inaction to effectively address health risks (Gadarian et al., 2020). The PEW Research Center acknowledged these dangers of public antipathy towards expertise when recognising that 'a scientific endeavour that is not trusted by the public cannot adequately contribute to society and will be diminished as a result' (Parikh, 2021). This is in clear recognition that collective action issues rely on public opinion which rightly serves to sway policy mechanisms and public responses.

A lack of trust can lead key decision-making officials, as well as the public, to question the 'facts' put forward by subject-matter experts (Cairney & Wellstead, 2020; Nichols, 2017), resulting in a perception of 'factual divergence', the term representing a move away from a level of scientific basis. This factual divergence is often thwarted by strong statements from scientists reasserting the expertise hierarchy when addressing misinformation (Calisher et al., 2020), or prompting the mere relay of 'clear, honest information to the public', otherwise known as the 'information deficit' model, an approach that many risk communication experts strongly critique (Mian & Khan, 2020).

Public risk perceptions, and subsequent outcries, have had a notable effect on government policy, despite the safety of the technology being asserted by experts. Risk is often communicated through expert bodies and governments to encourage change in public behaviours for perceived future or current hazards. This is not to presuppose that public swaying of policy is a negative outcome, but rather that a well-informed public is less susceptible to risk misconceptions, and hence will be more able to protest when protest is warranted, effecting change where it is not only justifiable but needed. Therefore, there is a constant feedback loop between public sentiment, governments and public policy regarding societal risk which inform individual risk perceptions.

As such, risk perceptions are covered by a myriad of factors that interplay to form what one 'sees'. The thing about what one 'sees' is that this is precisely the basis for action. Risk perception researchers will tell you that we can never truly know the risk, and going back to Socrates' and Theaetetus' conversation, even if we accurately measure the temperature, that measurement alone cannot dismiss someone's feeling of being cold in the wind, and consequently, their perception of chill.

References

Baumeister, R. F., Dewall, N., & Zhang, L. (2007). Do emotions improve or hinder the decision making process?. In K. D. Vohs, R. F. Baumeister, & G. F. Loewenstein (Eds.), *Do emotions help or hurt decision making: A Hedgefoxian perspective* (pp. 11–32). New York: Russell Sage Foundation.

Bodemer, N., Ruggeri, A., & Galesic, M. (2013). When dread risks are more dreadful than continuous risks: Comparing cumulative population losses over time. *PLoS ONE, 8*(6), 1–6. https://doi.org/10.1371/journal.pone.0066544

Cairney, P., & Wellstead, A. (2020). COVID-19: Effective policymaking depends on trust in experts, politicians, and the public. *Policy Design and Practice, 4*, 1–14. https://doi.org/10.1080/25741292.2020.1837466

Calisher, C., Carroll, D., Colwell, R., Corley, R. B., Daszak, P., Drosten, C., … Turner, M. (2020). Statement in support of the scientists, public health professionals, and medical professionals of China combatting COVID-19. *The Lancet, 395*(10226), e42–e43. https://doi.org/10.1016/s0140–6736(20)30418-9

Fischhoff, B. (1975). Hindsight not equal to foresight – Effect of outcome knowledge on judgment under uncertainty. *Journal of Experimental Psychology-Human Perception and Performance, 1*(3), 288–299. https://doi.org/10.1037//0096-1523.1.3.288

Fischhoff, B., Slovic, P., & Lichtenstein, S. (1983). "The Public" vs. "The Experts": Perceived vs. actual disagreements about risks of nuclear power. In V. Covello, G. Flamm, J. Rodricks, & R. Tardiff (Eds.), *The analysis of actual versus perceived risks* (pp. 235–249). New York and London: Plenum Press.

Fischhoff, B., Slovic, P., Lichtenstein, S., Read, S., & Combs, B. (1978). How safe is safe enough - psychometric study of attitudes towards technological risks and benefits. *Policy Sciences, 9*(2), 127–152. https://doi.org/10.1007/bf00143739

Gadarian, S. K., Goodman, S. W., & Pepinsky, T. B. (2020). Partisanship, health behavior, and policy attitudes in the early stages of the COVID-19 pandemic. *PloS One, 16*(4), e0249596. https://doi.org/10.1371/journal.pone.0249596

Gigerenzer, G. (2004). Dread risk, September 11, and fatal traffic accidents. *Psychological Science, 15*(4), 286.

Guy, S., Kashima, Y., Walker, I., & O'Neill, S. (2014). Investigating the effects of knowledge and ideology on climate change beliefs. *European Journal of Social Psychology, 44*(5), 421–429. https://doi.org/10.1002/ejsp.2039

Kahan, D. M., Peters, E., Wittlin, M., Slovic, P., Ouellette, L. L., Braman, D., & Mandel, G. (2012). The polarizing impact of science literacy and numeracy on perceived climate change risks. *Nature Climate Change, 2*, 732. https://doi.org/10.1038/nclimate1547

Keller, L. R., & Sarin, R. K. (1988). Equity in social risk: Some empirical observations. *Risk Analysis: An International Journal, 8*(1), 135.

Langer, E. (1982). The illusion of control. In D. Kahneman, P. Slovic, & A. Tversky (Eds.), *Judgment under uncertainty: Heuristics and biases* (pp. 231–238). Cambridge: Press Syndicate of the University of Cambridge.

Loewenstein, G. F., Weber, E. U., Hsee, C. K., & Welch, N. (2001). Risk as feelings. *Psychological Bulletin, 127*(2), 267–286. https://doi.org/10.1037//0033–2909.127.2.267

Mian, A., & Khan, S. (2020). Coronavirus: The spread of misinformation. *BMC Medicine, 18*(1), 1–2. https://doi.org/10.1186/s12916-020-01556-3

Nichols, T. M. (2017). *The death of expertise: The campaign against established knowledge and why it matters*. New York: Oxford University Press.

Parikh, S. (2021). Why we must rebuild trust in science. Retrieved from https://www.pewtrusts.org/en/trend/archive/winter-2021/why-we-must-rebuild-trust-in-science

Riechmann, D., & Madhani, A. (2020, April 27). No, don't inject disinfectant: Outcry over Trump's musing. *AP*. Retrieved from https://apnews.com/article/virus-outbreak-donald-trump-ap-top-news-politics-health-697d9ecef7f89cf5e9abb3b008c7faa7

Rollins, A. (2020, May 22). Govt 'sick of experts' in the public service: Leon, Brief article. *Canberra Times*. Retrieved from https://link.gale.com/apps/doc/A624541416/STND?u=unimelb&sid=STND&xid=e7a30dc9

Schlosberg, D., Collins, L. B., & Niemeyer, S. (2017). Adaptation policy and community discourse: Risk, vulnerability, and just transformation. *Environmental Politics, 26*(3), 413–437. https://doi.org/10.1080/09644016.2017.1287628

Schon, D. A. (1995). *The reflective practitioner: How professionals think in action*. London: Bookpoint Ltd.

Sedley, D. (2004). Knowledge is perception. In D. Sedley (Ed.), *The midwife of Platonism: Text and subtext in Plato's theaetetus* (pp. 38–53). Oxford: Oxford Academic. https://doi.org/10.1093/0199267030.003.0002

Sjöberg, L. (2000). Factors in risk perception. *Risk Analysis, 20*(1), 1–12. https://doi.org/10.1111/0272-4332.00001

Skjong, R., & Wentworth, B. H. (2001, June 17–22). *Expert judgment and risk perception*. Paper presented at the Eleventh International Offshore and Polar Engineering Conference, Stavanger, Norway.

Slovic, P. (1993). Perceived risk, trust, and democracy. *Risk Analysis: An International Journal, 675*, 675–682.

Slovic, P., Finucane, M. L., Peters, E., & MacGregor, D. G. (2004). Risk as analysis and risk as feelings: Some thoughts about affect, reason, risk, and rationality. *Risk Analysis, 24*(2), 311–322. https://doi.org/10.1111/j.0272–4332.2004.00433.x

Slovic, P., Fischhoff, B., & Lichtenstein, S. (1979). Rating the risks. *Environment, 21*(3), 14–20.

Slovic, P., Fischhoff, B., & Lichtenstein, S. (1981). Facts and fears: Societal perception of risk. *Advances in Consumer Research, 8*(1), 497–502.

Slovic, P., Fischhoff, B., & Lichtenstein, S. (1985). Characterising perceived risk. In R. W. Kates, C. Hohenemser, & J. X. Kasperson (Eds.), *Perilous progress: Managing the hazards of technology* (pp. 91–125). Boulder, CO: Westview.

Slovic, P., Peters, E., Grana, J., Berger, S., & Dieck, G. S. (2010). Risk perception of prescription drugs: Results of a national survey. In P. Slovic (Ed.), *The feeling of risk: New perspectives on risk perception* (pp. 261–284). London; Washington, DC: Earthscan.

Starr, C. (1969). Social benefit versus technological risk. *Science, 165*, 1232–1238.

Sunstein, C. R. (2003). Terrorism and probability neglect. *Journal of Risk and Uncertainty, 26*(2), 121–136. https://doi.org/10.1023/A:1024111006336

Taylor, S. E. (1982). The availability bias in social perception and interaction. In D. Kahneman, P. Slovic, & A. Tversky (Eds.), *Judgment under uncertainty: Heuristics and biases* (pp. 190–200). Cambridge: Cambridge University Press.

Tversky, A., & Kahneman, D. (1974). Judgment under uncertainty: Heuristics and biases. *Science*, 185(4157), 1124–1131

Wilson, R. S., Zwickle, A., & Walpole, H. (2019). Developing a broadly applicable measure of risk perception. *Risk Analysis, 39*(4), 777–791. https://doi.org/10.1111/risa.13207

Zinn, J. (2008). *Social theories of risk and uncertainty : An introduction*. Malden, MA: Blackwell Publishing.

4 Sociological Theories of Risk

"Dead are all the gods: now do we desire the Übermensch[1] to live" pronounced Zarathustra in Nietzsche's tale (Nietzsche, 2006, pp. Part I, Section XXII). This statement can be taken as a reflection of both the death of a belief system and also a liberation of it. It reflects on the potential for humans to step into their skin but also, I argue, to feel empowered to redefine risk in society. Risk is no longer in the realm of spirituality or higher powers as a tool to judge moral transgressions but one that is increasingly created and managed by humans. Being in charge of not only the onset of risks, but its definition, allows for a shifting of power from Gods to people (Foucault, 1982). This shift to human-centred risk has given rise to the increasing fascination in insurance and the reflexive modernisation that stems from such beguilement. This chapter discusses power aspects of risk definition, how it centres upon sociological cultural understandings of risk and also the way that risk is constructed within society to highlight our innermost fears in the context of new, manufactured risks. This chapter outlines key sociological theories of risk in ways that turn an introspective lens on ourselves to remind us that the Risk Society we are in is not one that has always been there. This chapter will also delve into Mary Douglas' work on cultural segments of risk, as well as touching on Governmentality.

Risks are always in the future. They are reflections of events that could, but may not, occur. As a result, they can be chopped, changed, altered and amplified in ways that may be a reflection of our own culture or society. The depictions of these risks, whether this be those discussed and importantly not discussed, provide a prism through which society's self-reflection can occur. This is the basis for Ulrich Beck's sociological approach (Beck, 1999, p. 73). The ontological basis for this position is that risk does not exist on its own merit, but rather is socially constructed to explain, reflect and speak for our society. Beck is a German sociologist most well known for his theory of the 'Risk Society', the idea that as a result of modernising, we have become overly obsessed with risks: with measuring it, with its management, with incessantly talking about it (Beck, 1992). This, Beck argues, is due to the 'new risks' that have taken shape following industrialisation. Beck's seminal work was written prior to the 1986 Chernobyl disaster, and released directly following

DOI: 10.4324/9781003432647-4

the event. As a result, the theory provided the much-needed societal lens that people were craving at that time. Chernobyl represented these 'new risks' in a way that resonated with readers instantly. These were risks that were deemed to be imperceptible, such as radiation spreading across Europe from its Eastern regions. Imperceptibility implies that it is not possible to discern whether one is exposed to a hazard purely on the basis of human perception. Therefore, this 'new risk' is difficult to *feel.* Other characteristics of new risks include the indiscriminate nature of these risks, i.e. that everyone is affected by it, its global scope transcending borders in ways not seen before industrialisation. Finally and importantly, these risks are manufactured through human endeavours. Often seen as 'side effects' of modernisation processes, these manufactured risks lie in direct contrast to traditional risks that are rooted in natural determinants (such as floods and tornadoes). As these new risks have taken shape, Beck argues that society has increasingly become obsessed with their management. He has coined this the *Risk Society*.

In a *Risk Society*, we see the increasing emergence of manufactured risks around us leading societies to reflect deeply in their role in the creation of such risks, and relating this to their technological, economic and social trajectories. This leads us to acknowledge the growing role of uncertainty in modernisation processes, such as environmental degradation and technological advancements. In response, we recognise the risks that are inherent in such processes, and seek to eradicate them through increasing technological change. This, Beck argues, in turn creates a never-ending cycle of risks. Technological change brings risks in its use, which is then offset by the creation of new technology, which creates its very own risks. Beck (2006) distinguishes a clear difference in the risks we face today as being nothing like anything previously seen in human history. This is particularly what makes our current epoch so unique: widespread environmental devastation, radiation impacts and ecosystem challenges. The scale of these risks is global, and the ability to address them has become increasingly challenging due to the organised irresponsibility of causal actors. They are incalculable and importantly non-insurable. Their effects are long-lasting and irreparable to the point where compensation is not possible (Lupton, 2013). The indiscriminate nature of such risks lends itself to a sense of democracy: "poverty is hierarchic, but smog is democratic" (Beck, 2009). Contamination in a water source, for example, does not stop at a wealthy person's tap.

In a critical structuralist imagining, Beck recognises the importance of risk definition. This is a power afforded to specific groups, often experts, to judge on behalf of all as to the acceptable level of risk that society can face. This power is derived from not only the institutionalised norms and positioning but also the knowledge provided to, and by, such groups. In this way, risk has become a politically contested zone. In Foucault's writings, power and knowledge are inextricably intertwined. There is power in the attainment of knowledge, and in turn those with power can define what is considered 'truth' (Foucault, 1982). Risk is, after all, a process of knowledge and social construction that is constantly negotiated and re-negotiated among people. Knowledge, in such situations, cannot be 'unbiased' or 'neutral', but must always carry the implication of interest in its development (Lupton, 2013, p. 39). "[A] risk, therefore, is not a static, objective phenomenon, but is constantly constructed … as part of the work of social interaction and… meaning" (Lupton, 2013, p. 44).

Beck sees the placing of knowledge and power in 'experts' as problematic. In particular, the infantilisation of laypeople to unequivocally accept expert judgments creates a worrying two-tier society that centralises the power of risk definition into one group (experts). It provides technical fields with a monopoly of defining, and measuring risks that often allow little discussion on risk tolerance in society more generally. Acceptance of risk, argues Beck, is not one that can be built purely off science. It is a social question, and is further implicated by the 'seeds of doubt' (Beck, 2009, p. 35) that underlie previous scientific declarations of risk. There have been a number of public debates on these very notions playing out, particularly on the impacts of new technologies. Experts have the power to frame and position risks in society but are as fallible as a layperson in terms of the influence of social practices and attention cycles (Lupton, 2013). On this point of fallibility is where we see strong alignment between psychological discourses of risk and sociological understandings.

Governmentality

Where Beck's Risk Society thesis takes a path from being critical realist through to more constructivist angles over time, Foucault and Lupton's work on Governmentality firmly has its roots in strong social constructivism. Experts, Lupton contends, still relate socially and culturally within society and as such, cannot have the unbiased perspective that is often touted in technocratic circles (Lupton, 2013). Governmentality approaches to risk take Foucault's knowledge/power nexus work further to encompass the surveillance and regulation of populations to encourage self-risk mitigation activities. The purpose of governmentality is to direct human behaviour. Underpinned by notions of liberalism, Governmentality is a way of ensuring levels of compliance, without the direct means imposed by a 'large' government.

For example, Governmentality has been used in promoting risk management behaviours in potable water intake. Across South-Eastern Australia in 2021 and 2022, El Nino weather patterns had resulted in the inundation of much of the region. This had a dire effect on drinking water in the network, prompting local authorities to issue Boil Water Alerts. These alerts ask residents to first boil (and cool) their drinking water prior to consuming. This was as a result of high infiltration of bacteria, algae or other operational issues that rendered the water unsafe to drink from the tap. The warnings are distributed using a variety of media, letter drops and direct contact with residents. Residents are asked to manage their own risk by way of changing their behaviour.

Similarly, Governmentality methods have been used to change social norms during the Millennium Drought in Australia. During the height of water restrictions in Melbourne, residents were prompted not to use potable water directly on gardens. Green, lush gardens without clear 'bore water/tankwater in use' signs were looked at in disdain by neighbours. The campaigns led by the water authorities and the Victorian state government successfully created norms that developed social pressure within communities who utilised surveillance across their neighbourhoods. Foucault's panopticon was in full swing. Communities were regularly monitoring each other, surveillance was at an all-time high, and neighbours were reporting illegal garden watering to the local authorities for action. In this way, Governmentality was used to regulate individual risk behaviour in ways that became culturally self-enforcing.

A Cultural Risk Approach: Grid-Group Typology and Risk

Agents are undeniably embedded in a particular culture that drives risk. This culture is one that is made up of norms and expectations, and as defined by Mary Douglas "the publicly shared collection of principles and values used at any one time to justify behaviour" (Douglas, 2010, p. 67). A person is surrounded by the invisible pull of cultural understandings and worldviews, and these form the basis for decision-making. The anthropological basis of the Cultural Risk theory is not new, but rather has existed for the better part of the last century. Attempting to explain, and predict, risk understandings through membership of a particular worldview, it captured the attention of risk researchers for its ability to incorporate meso-level groupings based on values and beliefs (Koehler, Rayner, Katuva, Thomson, & Hope, 2018).

Mary Douglas and Aaron Wildavsky (1982) argue that the grid-group segment, or worldview that one prescribes to, ultimately alters the way that an individual may perceive a risk. A worldview encompasses the way that people view the hierarchy of systems that they are placed in, and the amount of autonomy afforded to them as a result. This cultural theory sees individuals as part of a conglomerate sharing similar worldviews.

'Hierarchy' and 'Markets' were previously the two main antithetical groupings used in the social sciences to cluster similar cultural worldviews (Thompson, Ellis, & Wildavsky, 1990). If prescribing to a market typology, the individual has more freedom to explore new approaches to the way they choose to engage in life situations rather than one who subscribes to a hierarchical view. The ensuing result is based on the individual's own actions, hence why it is considered the 'market-based' approach (think Liberalism). If adhering to the 'Hierarchy' line, the individual is heavily captured by their social standing in determining how much influence they have over themselves and those around them. These two main groupings formed the basis for many economic and social studies in the 1970s (Lindblom, 1977; Williamson, 1975). As an anthropologist, Douglas (2013) determined that there were social groupings missing in this set, and sought to create a more robust clustering of worldviews: the grid-group typology.

The grid-group typology had been designed by Douglas to show the 'variability of an individual's involvement in social life' and is captured by two distinct factors: grid and group. The level of individual social control is the key emphasis in the underlying nature of a grid-group approach. They found that not only did it matter how people were grouped into associations and consortiums, but also whether these groupings were structured internally or not. Without a grouping, a social structure can still exist. This created an arrangement similar to a Cartesian plane, made up of grid on one axis with group on the other (see Figure 4.1). Mary Douglas and Aaron Wildavsky (another anthropologist) (1982) argue that the grid-group segment, or worldview that one prescribes to, ultimately alters the way that an individual may perceive a risk. A worldview encompasses the way that people view that hierarchy of systems should be structured, and the amount of autonomy afforded to them as a result. This theory thus sees individuals as part of a conglomerate sharing similar worldviews.

Group and Grid Explained

On the group axis (refer to Figure 4.1), the extent to which an individual's life is absorbed and influenced by group membership is considered (Douglas, 1992). This is an important element in understanding the boundaries of a group, how inclusive their membership may be and also how an individual's own perceptions could be altered by their involvement in the group. A low group membership, or low group on the plane, implies a higher level of individuality of a person, and also a higher level of control over their own affairs and resources. Douglas (1996, p. viii) defines group as 'the experience of a bounded social unit' highlighting that a high group is about the group itself rather than the individuals. A high group will be defined by a stark contrast between 'members' and 'non-members', with members being afforded the right to shared resources with others within their group.

Grid, as distinct from Group (Figure 4.1), describes the level of social structure between people (Douglas, 1992, 2013), essentially defining how individuals relate to one another. Grid can refer to the level of individual power over others. In a low grid, people are less defined by a particular role and their individual social status, implying that the networks between people are relatively the same, regardless of the individual that you may approach. Conversely, a high grid dimension carries clearly defined social regulations that vary based on the individual, for example, military hierarchy or caste societies. Decisions in a high grid scenario are made on behalf of many by a few

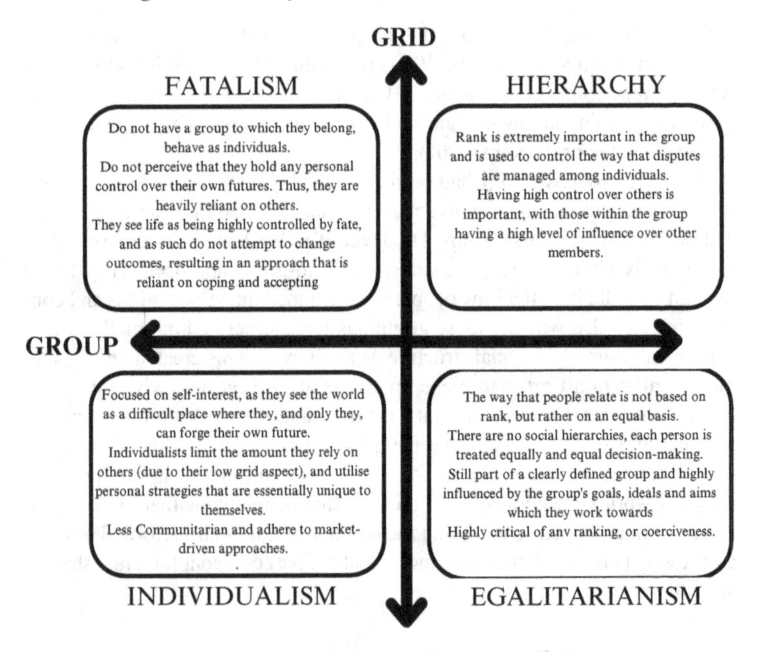

Figure 4.1 Grid-group scales and their segments.

'elites'. Networks between individuals carry different weightings and levels of power over one another depending on where the individuals are placed on the social hierarchy. These relationships may shift and change based on existing status frameworks. High grid implies that a socially imposed structure exists between people that may limit what they can and cannot do, based on their social status or level within the structure.

The Cartesian plane made up of grid and group forms four distinct social groupings: Hierarchy, Egalitarianism, Individualism and Fatalism (Figure 4.1).

High Group/High Grid = Hierarchy
High Group/Low Grid = Egalitarianism
Low Group/Low Grid = Individualism
Low Group/High Grid = Fatalism

Finally, the literature also explores a fifth group, 'Hermit', as a classification for those individuals that do not fit into any of the above four categories. This group sits outside of the cultural plane, as these individuals are relatively unconcerned with any social structures or make-ups. They actively disengage from social contribution (Thompson et al., 1990).

Hierarchy

The hierarchy dimension exists in the high group and high grid of the plane. This group has clear exclusive membership, with clearly defined boundaries around its structure. The high grid nature of the hierarchy grouping suggests a rigidly defined structure of the social makeup of the group. Although a group exists, each individual is ranked, unequally, based on their social status.

A member of a hierarchy is within a clear, bounded group with many rules and regulations that stipulate its membership. Rank is extremely important in the group and is used to control the way that disputes are managed among individuals. Having high control over others is important, with a select few within the group holding a high level of influence over other members. Thompson et al. (1990) use the example of a hierarchy in a high-caste Hindu village. This villager is within a clear, bounded group, and is ranked relatively highly within their sphere. This affords them a substantial level of control over others based on their position in the group. They also must adhere to the regulations and rules stipulated by their membership. In this way, their freedom is 'constrained by socially imposed gridiron' (Thompson et al., 1990, p. 7).

Egalitarianism

Egalitarianism carries the same group influences and membership as the hierarchy dimension; however, it differs through its internal social structures. The way that one person relates to another is not based on rank or hierarchy, but rather on an equal basis. Egalitarianism provides very little basis for excluding members, unlike the hierarchy faction. Networks between people within the Egalitarianism group do not change from person to person, a distinction from the rigid ranking structure of a hierarchy. Unsurprisingly, Egalitarianism focuses on the equal footing of each person within the group, with the desire to abolish coerciveness and inequality (Thompson et al., 1990).

A key example of Egalitarianism is a member of a commune. There are no social hierarchies to speak of; each person is treated equally and is able to equally make their own decisions. They are still part of a clearly defined group and highly influenced by the group's goals, ideals and aims which they work collectively towards. The group is highly critical of any ranking or coerciveness, which can then prove difficult to resolve disputes. Thompson et al. (1990) highlight that Egalitarians see themselves as effectively having a 'wall of virtue' that outlines and bounds their group.

Individualism

Individualism is housed under the low grid, low group segment. This is due to those in the group coveting freedom from coercion and the influence from others, while also not belonging to any particular faction. They are predominantly focused on self-interest, as they see the world as a difficult place where they, and only them, can forge their own future. In this way, individualists limit the amount they rely on others (due to their low grid aspect), and utilise personal strategies that are essentially unique to themselves (Thompson et al., 1990).

An example of an individualist is a self-made businessperson, perhaps someone who has established their business on their own. They utilise market-based approaches to get ahead, with self-interest acting as the key motivator to this behaviour. This is a starkly different approach to the other three groups, who either have some element of socially imposed gridiron of rankings or are part of a group with a common goal.

In the case of Australia and the Murray Darling Basin, water rights are bought and sold on the market, allowing individuals and private entities the right to extract water when needed. This type of approach adheres to the Individualist worldview, as it allows people to be able to take control over how much water they use. Oftentimes, government policies (such as in Australia) would favour the rights of individual landowners or corporations to extract or use water for their own purposes, potentially leading to over-extraction and ensuing environmental waterway degradation. The emphasis is on the freedom of individuals to use water as they see fit, with minimal government intervention.

This individualistic mindset in water governance can sometimes create conflicts between different stakeholders, especially when the actions of individuals or corporations affect the water supply for neighbouring communities, agriculture or the environment. Balancing individual rights with the collective need for sustainable water management remains a significant challenge for many governments worldwide.

Fatalism

The final quadrant of the plane describes the low-group, highly controlled social imposition of the fatalists. They ultimately do not have a group to which they belong, and as such behave as individuals. The difference between this and individualism is that fatalists do not perceive that they hold any personal control over their own futures. Thus, they are heavily reliant on others. They see life as being heavily controlled by a 'higher power', and as such do not attempt to change outcomes, resulting in an approach that is reliant on coping and accepting (Thompson et al., 1990). Luck, fortune and lottery define this group, resulting in a viewpoint that accepts outcomes readily. An example put forward by Thompson et al. (1990) is a non-unionised factory worker. This factory worker's life can be seen to be heavily controlled by others: their bosses, the government and also their own surroundings. They are resigned to the fact that these entities will have more control over their personal lives than they themselves do. Their worldviews then reflect this approach, as they do not join a union, and do not attempt to change elements of their lives.

Grid-group Membership and Risk

The worldviews described: Hierarchy, Egalitarianism, Individualism and Fatalism, all are theorised to affect the way in which an individual treats risks. A fatalist may see risk as something that cannot be shifted or altered according to their own actions. That risk is essentially 'there' and 'not controllable'. They would accept decisions made by others in terms of risk acceptability and feel that they have no power in whether they are exposed to the risk or not. In this way, a fatalist may respond differently to a risk than an individualist. The individualist conversely sees risk as something that can be controlled to some extent, but all decisions relating to its exposure should be made by the exposed individuals themselves, not by any other entity or group. For example, an individualist may see the choice to fluoridate their water as a right of their own. Therefore, the risk that is placed on them should only be as a result of their own decisions, not decisions made by others being imposed upon them. Their worldview encompasses and is heavily based on the importance of personal choice.

An Egalitarian instead focuses on the common good rather than that of an individual's self-interest. For example, in the case of an Egalitarian's view of risk the individual risk of fluoridating water is more acceptable when it means net benefit at the public health level. The risk of a person receiving an adverse reaction to fluoride is considered incredibly small when compared to the overall benefits to society. In addition to this, the decision-making power of risk acceptability also lies with the group, not with a select few (highlighting the low grid approach). This can mean that, in an Egalitarian mindset, public perception can make a large impact upon risk-based decision-making, as essentially, each vote would be treated equally, regardless of expertise.

The hierarchy faction is also based on adhering to decision-making on behalf of the whole group (can be also seen as 'big' government). However, the key difference exists in how the decision on risk acceptability is made. In the case of Egalitarianism, this decision is made by the whole group, in a democratic process, with each member representing an equal proportion of the vote. Within the hierarchy quadrant, it is the decision of a few that outline the approach for the group. In the case of converting a water supply from dam water to potable recycled water, a group of experts would highlight their risk acceptability, when it comes to whether to proceed or not. In a hierarchical worldview, this decision should not be decided by many, but by those few that are granted this right on the basis of their expertise. The decision-making power for risk acceptability then does not fall to the public, but rather to a few. This can be seen as an advantage, as quite often these 'few' are chosen based on their position within the hierarchy, or have some expertise within the field. However, a low grid adherer may begrudge the lack of personal autonomy in the process. The individualist would want to make the decision themselves, and only themselves, affording full autonomy with whether to proceed, based on self-interest. Egalitarians would begrudge the hierarchy for ignoring the risk perceptions of the layperson and treating people who must be exposed to the risk as essentially having no say over their own personal exposure.

As each worldview considers risk and personal risk acceptance in differing ways, tensions arise when people of different worldviews interact. Hierarchists would always defer to experts, whereas an Egalitarian would seek public consultation or public opinion on risk acceptability. Fatalists would take little, if any, action on the decision-making process of a risk. Individualists would like very little decision-making that is group based, but rather desire full autonomy over their own risk exposure. Thus, the worldview of the person undertaking the risk would greatly affect the process by which they decide, or defer decisions, on risk acceptability. Therefore, one could argue that the decision of a Hierarchist on a particular issue or project could vary from the decision from someone who sees themselves as an individualist. This inconsistency in decisions ultimately changes whether a project proceeds or not.

What Empirical Evidence Underlies these Theories?

Many sociologists or anthropologists would argue that the attempt to 'measure' cultural worldviews or risk societies is a doomed enterprise; instead, research methods such as ethnographic research are more suited to studies of culture and society. Nevertheless, there were studies that attempted to measure these theories to see if they would stack up quantitatively. In gathering empirical evidence, a study by Palmer (1996) sought to determine whether risk perceptions were linked to cultural worldviews, resulting in mixed findings. Her methodology did well to predict and link risk perceptions of Hierarchists, yet generated varied results for Egalitarians and Individualists. There was no testing of Fatalists. Similarly, others did an extensive comparison of risk theories, and their empirical bases, concluding that the cultural theory had not been able to explain more than 5–10% of the variance in perceived risk (Sjoberg, 1998; Sjöberg, 2000). Similarly, Marris, Langford and O'Riordan (1998) also showed little correlation between cultural biases and risk perceptions. It is important to note that these articles are now quite dated.

And then there were the studies that had positive results. Perrella and Kiss (2015) sought to measure the extent to which cultural dimensions could impact on positions on fluoride in drinking water. Their Canadian study confirmed that membership to the Egalitarian quadrant of the worldview matrix indicated the commitment to group decision-making, and a general sense of wanting to decrease societal inequity, both factors that resulted in a pro-fluoridation view. The United States, however, is a culture that exhibits Individualistic tendencies, which drives much of the governance and societal decision-making. Individualists, unsurprisingly, wanted to have their own water supply rather than be part of a city's water grid. This ensures independence and increased amount of autonomy over their livelihoods. Even when faced with a contaminated water supply, there was still reluctance on the part of Individualists to connect to the clean reticulated system (Fitchen, 1990). The value placed on self-reliance was the overriding factor in shaping risk perceptions regarding a water contamination scenario. So the verdict is still out whether cultural theory can stand up to quantitative scrutiny. Similarly, the Risk Society thesis also exists on shaky empirical footing (Rasborg, 2012). It is conceptual and relies on inferences for its basis, which is not to say that it is less valuable than an overly quantified notion of risk. There has been criticism posed at the western-centric nature of the concept and its ability to explain a global phenomenon that is not euro-centric. Furthermore, the process of self-reflection and individuality put forward is one that is not socio-economically viable for many, and thus is only within the reach of elites (Lash, 1994). The cultural and material resources required to undergo such introspection are often not available to many, and as such have been criticised for acting as a distinction between "reflexivity winners and… losers" (Lupton, 2013, p. 156).

After exploring various risk factors, the upcoming chapter will introduce a study focused on water decision-makers. This study aims to shed light on how different views of risk influence decision-making processes, offering valuable insights into the factors that sway decisions.

Note

1 Übermensch has been translated as Superman (Thomas Common translation) and refers the ideal that humans strive towards. Übermensch does not presuppose that humans can take the place of God, but rather to allow humans to gain command over their futures (Lemm, 2021).

References

Beck, U. (1992). *Risk society: Towards a new modernity*. London: Sage Publications.

Beck, U. (1999). *World risk society*. Malden, MA: Polity Press.

Beck, U. (2006). Living in the world risk society: A Hobhouse Memorial Public Lecture given on Wednesday 15 February 2006 at the London School of Economics. *Economy and Society, 35*(3), 329.

Beck, U. (2009). *World at risk*. Malden, MA: Polity Press.

Douglas, M. (1992). *Risk and blame: Essays in cultural theory*. London: Routledge.

Douglas, M. (1996). *Natural symbols [electronic resource]: Explorations in cosmology*. London; New York: Routledge.

Douglas, M. (2010). *Risk acceptability according to the social sciences* (1st paperback ed. ed.). London: Routledge.

Douglas, M. (2013). *Essays in the sociology of perception*. London: Routledge.

Douglas, M., & Wildavsky, A. B. (1982). *Risk and culture [electronic resource]: An essay on the selection of technological and environmental dangers*. Berkeley: University of California Press (1983 printing).

Fitchen, J. M. (1990). Cultural values affecting risk perception: Individualism and the perception of toxicological risks. In L. A. Cox & P. F. Ricci (Eds.), *New risks: Issues and management* (pp. 599–607). Boston, MA: Springer US.

Foucault, M. (1982). The subject and power. *Critical Inquiry, 8*(4), 777–795. https://doi.org/10.1086/448181

Koehler, J., Rayner, S., Katuva, J., Thomson, P., & Hope, R. (2018). A cultural theory of drinking water risks, values and institutional change. *Global Environmental Change, 50*, 268–277. https://doi.org/10.1016/j.gloenvcha.2018.03.006

Lash, S. (1994). Expert-systems or situated interpretation? Culture and institutions in disorganized capitalism. In U. Beck, A. Giddens & S. Lash (Eds.), *Reflexive modernization: Politics, tradition and aesthetics in the modern social order* (pp. 198–215). Stanford: Stanford University Press

Lemm, V. (2021). "5. The Work of Art and the Death of God in Nietzsche and Agamben". In M. Norris & C. Dickinson (Eds.), *Agamben and the Existentialists* (pp. 83-99). Edinburgh: Edinburgh University Press. https://doi.org/10.1515/9781474478793-006

Lindblom, C. E. (1977). *Politics and markets: The world's political-economic systems.* New York: Basic Books.

Lupton, D. (2013). *Risk* (2nd ed.). New York and Oxon: Routledge.

Marris, C., Langford, I., & O'Riordan, T. (1998). A quantitative test of the cultural theory of risk perceptions: Comparison with the psychometric paradigm. *Risk Analysis, 18*(5), 635–647.

Nietzsche, F. W. (2006). *Thus spoke Zarathustra: A book for all and none* (T. Cannon, Trans.). Cambridge: Cambridge University Press.

Palmer, C. G. S. (1996). Risk perception: An empirical study of the relationship between worldview and the risk construct. *Risk Analysis, 16*(5), 717–723. https://doi.org/10.1111/j.1539-6924.1996.tb00820.x

Perrella, A. M. L., & Kiss, S. J. (2015). Risk perception, psychological heuristics and the water fluoridation controversy. *Canadian Journal of Public Health, 106*(4), e197–e203. https://doi.org/10.17269/cjph.106.4828

Rasborg, K. (2012). '(World) risk society' or 'new rationalities of risk'? A critical discussion of Ulrich Beck's theory of reflexive modernity. *Thesis Eleven, 108*(1), 3–25. https://doi.org/10.1177/0725513611421479

Sjoberg, L. (1998). Explaining risk perception: An empirical evaluation in cultural theory. In R. Lofstedt & L. J. Frewer (Eds.), *The Earthscan reader in risk and modern society* (Vol. 2, pp. 115–132). London: Earthscan.

Sjöberg, L. (2000). Factors in risk perception. *Risk Analysis, 20*(1), 1–12. https://doi.org/10.1111/0272-4332.00001

Thompson, M., Ellis, R., & Wildavsky, A. (1990). *Cultural theory*. Boulder, CO: Westview Press.

Williamson, O. E. (1975). *Markets and hierarchies, analysis and antitrust implications: A study in the economics of internal organization*. New York: Free Press.

5 Water Practitioners

How Do They See Risk? A Study

A water professional walks into their office on a typical day, to sit down at their desk to figure out the best option for a supply problem. They have three options in front of them, all different in their own way. One option is the 'business as usual', which means continuing to do what has been done in the past: fixing and maintaining of pipes. The second option considers a local community campaign to reduce demand on water use. The third option incorporates new technology that has never been used before to augment supply. What will the person do? Probably unsurprisingly, it all depends on whose desk this project lands on in any given day. Mohamed will rank it different from Maddy who will in turn score differently from Jing. Does this surprise you? Or will you adamantly cry out that there are processes in place to manage such deviations in expert opinion?

I put forward this question to a group of water professionals across four different organisations in Australia. I sought to test whether our processes – namely, risk matrices – are sufficient in alleviating the varied distinctions between assessors. Some argued that yes, the processes are flawed, but what else have we got? Others staunchly pointed to the risk matrix guide and highlighted the specificity of scoring metrics attached to each risk topic. I sought to uncover what was actually happening in practice.

The risk measurement approaches within water industries globally are predominantly based on existing international standards for risk assessments. The standards adopt a theory of assessment that is grounded in the technical risk approach, a theory which was conceived in the 1950s and 1960s which stem from millennia of quantification (see Chapter 1). The technical approach centres on the theory that risk exists, and that can be measured objectively. The 'rational' risk assessor underpins the basis of the theory, which leads to results that are objective and 'true'. The probabilistic risk assessment framework is a form of the technical risk approach that is applied in risk assessments throughout the world. This framework quantifies risk through a method of first determining its likelihood (or probability) of occurring to the magnitude of the consequence of the hazard. The two figures are used to determine an overall risk score, with the use of a 'risk matrix', a table of risk ratings that

DOI: 10.4324/9781003432647-5

have two inputs: likelihood and consequence. As one of the most notable uses of a risk matrix, the US Nuclear Reactor Agency report on Reactor Safety published in 1975 paved the way for global adoption of the tool, cementing its place in risk assessments worldwide (see Chapter 2).

The Matrix – Is it On Its Way Out?

Chapter 1 and 2 described the enduring quality of risk matrices in our society. Starting with Arnobius, in the 4th century AD (Simmons, 1995), risk matrices have not changed significantly over time, and still retain the basic principles of the initial idea by Arnobius:

Arnobius questions the different pathways that one can take with regard to religiosity, and highlights the potential benefits and losses from them (Table 5.1):

> Since, then, the nature of the future is such that it cannot be grasped and comprehended by any anticipation, is it not more rational, of two things uncertain and hanging in doubtful suspense, rather to believe that which carries with it some hopes, than that which brings none at all? For in the one case there is no danger, if that which is said to be at hand should prove vain and groundless; in the other there is the greatest loss, even the loss of salvation, if, when the time has come, it be shown that there was nothing false in what was declared.
>
> (Franklin, 2015, pp. 249–250)

Pascal revisits this question many centuries later (dubbed 'Pascal's Wager') and reassesses it with probability theorems (Neiva, 2023). Interestingly, Pascal provided this assessment as a way of showing that some questions cannot be answered with logical reasoning, namely, those that relate to the existence of God (Pascal, 1958 (1670)). Despite this, the way that we have been assessing risk arguably has not changed for millennia, and it used to inform all manner of topics (although no longer used for religious justifications!). Some may say that it is enduring because of its ease of use and its ability to convey information to inform decision-making. A key element of the staying-power of risk matrices exists in its simplicity. As a tool that can be rolled out across an organisation with diverse functions, while also allowing for a seemingly rational

Table 5.1 Adapted view of *Pascal's Wager*, the matrix for entry into the afterlife

	To believe and 'do good'	*To not believe and 'not do good'*
God exists	Entry to afterlife, heaven (Infinity rewards)	Denied entry to afterlife
God does not exist	No difference	No difference

approach of quantifying risk, it is a tempting template to use. However, behind these risk measurement tools exists degrees of subjectivity.

The technical risk assessment's reliance on the rational actor is predicated on the belief that it will produce decisions that are positivist and objective. The assumption stands that regardless of the risk assessor, if they are rational, it should always result in the same positivist outcome. Much of the technical-scientific literature focuses on the issues of identification of risk, how this is calculated, together with the accurate nature of it. In the last two decades, many theorists have presented criticisms of the technical risk measurement. In particular, this approach cannot easily quantify social effects of risks, as it relies heavily on the availability of legitimate statistical and fiscal data.

Risk assessments, in practice, diverge from the leaps and bounds made in the academic literature. Research in the past decade has described the impracticality and dangerous actions of utilising risk matrices to measure hazards (see Chapter 2).

Many other risk theories also exist that refute this approach, such as those based on the psychological, sociological and cultural aspects while also providing alternative viewpoints of the risk debate, as described in Chapters 3 and 4. Psychological approaches to risk, such as the effect of cognitive bias on risk perceptions, have propagated within the literature (Chapter 3). Furthermore, sociological theories in 'new risks', and also cultural theories, have taken a stand in explaining risk and risk behaviours (Chapter 4). These are not commonly reflected in current uses of risk measurement approaches.

Other studies have been undertaken to show the language and rhetoric of risk in the water industry, these varying from many differing types of risk, such as reputation-based risk and safety-based risk; however, a quantitative assessment on this process has not been yet undertaken. The research explored within this study considers the nature of existing risk assessments, and whether they can, in fact, be considered objective.

I provided water professionals across four water authorities with seven fictional projects that needed risk scores. Each respondent provided risk scores from 0 to 25 using their own risk organisational risk scoring process (these were adapted to ensure comparability between organisations).

The seven projects provided were:

1 Replacement of 400 mm water pipe in a busy residential area
2 Pump Station Installation
3 Construction of Recycled Water Treatment Plant
4 Save Water Campaign
5 Using Recycled Water as Potable
6 Using Radiation in Treatment of Drinking Water
7 Removing Fluoride from Drinking Water Supply

Table 5.2 A typical matrix of water authorities in the study (F. Portelli, personal communication, 26 June 2018)

Consequence/ likelihood	*1 (Low consequence)*	*2*	*3*	*4*	*5 (High consequence)*
1 (Very unlikely)	1 (Low)	2 (Low)	3 (Low)	4 (Low)	5 (Low)
2	2 (Low)	4 (Low)	6 (Med)	8 (Med)	10 (Med)
3	3 (Low)	6 (Med)	9 (Med)	12 (High)	15 (High)
4	4 (Low)	8 (Med)	12 (High)	16 (High)	20 (Extr)
5 (Highly likely)	5 (Low)	10 (Med)	15 (High)	20 (Extr)	25 (Extr)

The risk score was formulated from two separate assessments: risk likelihood and risk consequence. Both scores ranged from 0 to 5, and were then multiplied to form the final risk score. There were also four risk ratings: low (1–5), medium (6–10), high (11–16) and extreme (16+). When reporting on risk, these authorities typically report the risk score using the risk matrix shown in Table 5.2, hence why this one-dimensional figure is utilised in this study.

Defining 'Objective'

Objectivity in this context assumes that each risk assessor has the same risk assessment outcome when presented with identical project information. The process was deemed as being objective if respondents reported scores that fell within the same risk rating (low, medium, high or extreme). All participants were provided with their organisational predefined risk assessment procedure, and this was used to determine their risk score. The risk rating is ultimately one of the major decision-making mechanisms in funding allocations and options assessments within many segments of the water industry; therefore, provided a score is within the same category, it can be said to not affect the outcome of the progression of a project drastically. This process tests the impact of the individual upon risk assessments and therefore, any subjectivity that may arise from personal risk perceptions.

The risk categories are outlined in Table 5.2. Please note that due to the multiplication of consequence and likelihood scores (both out of 5), there were some risk scores that could not be obtained. For example, 17–19 were excluded from the table as they were not possible through the multiplication of two numbers between 1 and 5.

Note: As the scores were made up of a multiplication of two factors, consequence and likelihood, it created a statistical anomaly that rendered the data more likely to be positively skewed. The data was taken as the square root of each figure, as this was conceptually appealing due to the multiplication of consequence and the risk equating to the geometric mean of the two (Table 5.3).

In the assessment of individual risk ratings of water professionals (Figures 5.1 and 5.2), the study exhibited a fair amount of variation between individuals. Each risk assessor was provided with the same information on each project as well as an identical risk assessment process for their organisation and yet the scores varied wildly between each respondent. The first comparison was assessing Projects 1–4. These projects were all considered 'familiar' and 'business as usual' within the water sector, generally. As such, they were projects that each individual would have undertaken similar assessments for in the past (Table 5.4).

The difference between the projects themselves became apparent when considering the range within one standard deviation. Approximately 68% of the data fell within the range of low to high risk, whereas the same amount of data fell between medium and extreme for Project 3. The data continued to disperse even further when considering the range within two standard

Table 5.3 Risk rating score range

Risk rating	*Range from (inclusive)*	*Range to (inclusive)*
Low	1	4
Medium	5	9
High	10	16
Extreme	20	25

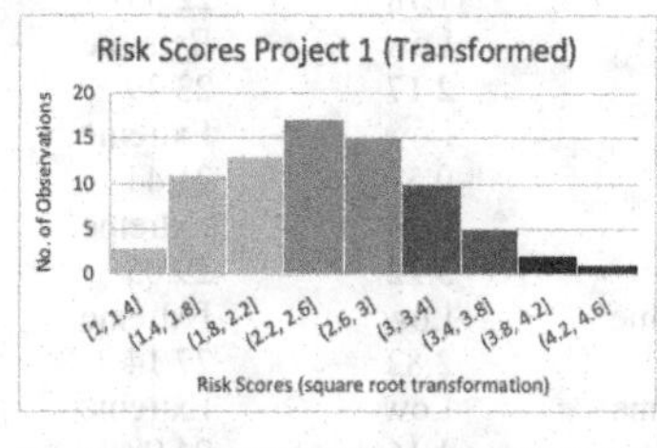

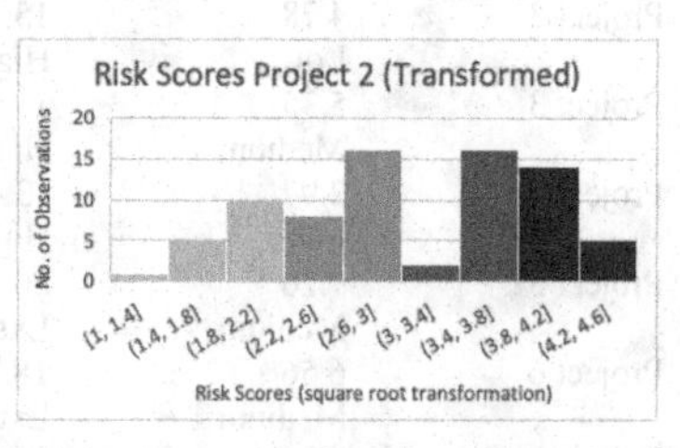

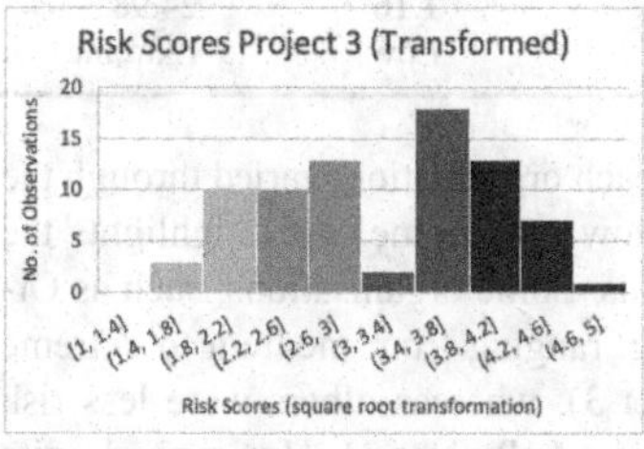

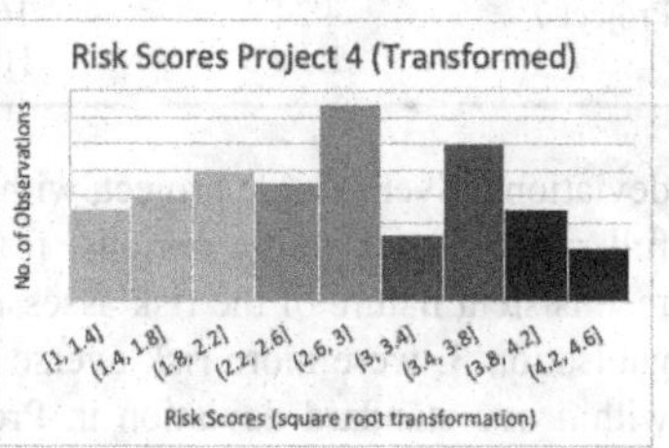

Figure 5.1 Histogram of risk scores (Projects 1–4). Light grey represents Low Risk, while the darkest grey represents extreme risk.

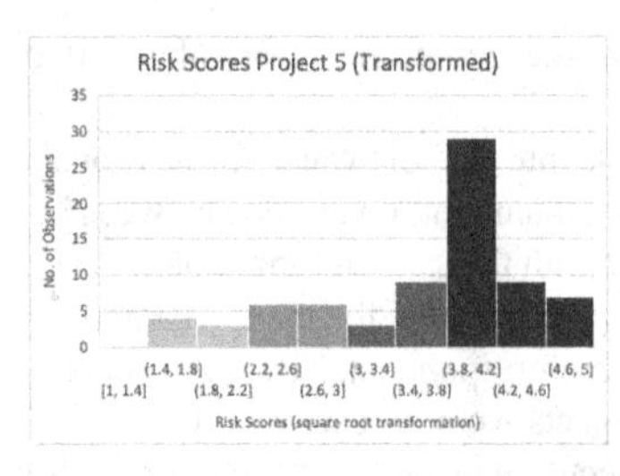

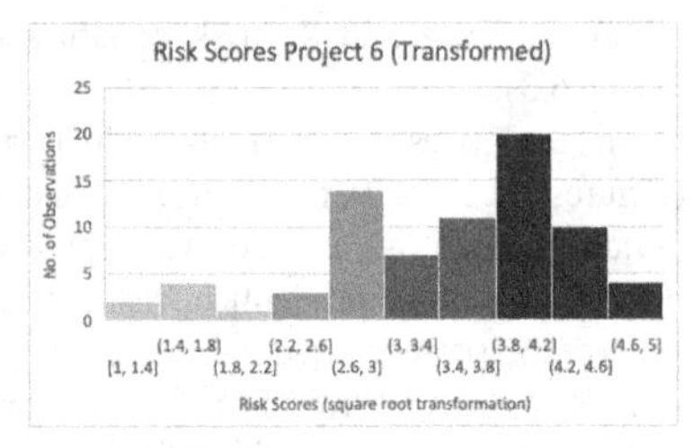

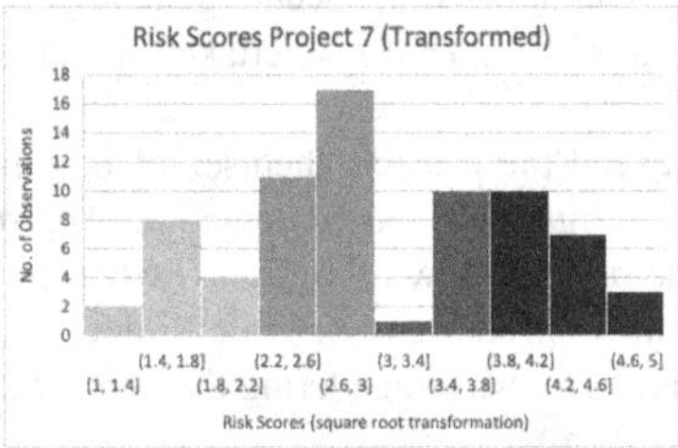

Figure 5.2 Histogram of risk scores (Projects 5–7). Light grey represents Low Risk, while the darkest grey represents extreme risk.

Table 5.4 Scores by project (back-transformed) within one and two standard deviations

Project no.	*Range within 1 SD (~68%)*		*Range within 2 SD (~95%)*	
Project 1	3.16	10.33	1.12	15.44
	Low	High	Low	High
Project 2	4.78	15.16	1.78	22.53
	Low	High	Low	Extreme
Project 3	5.35	15.96	2.17	23.39
	Medium	High	Low	Extreme
Project 4	2.97	13.41	0.57	21.44
	Low	High	Low	Extreme
Project 5	7.26	20.16	3.22	29.03
	Medium	Extreme	Low	Extreme
Project 6	6.56	18.72	2.82	27.14
	Medium	Extreme	Low	Extreme
Project 7	4.24	16.14	1.16	24.98
	Low	High	Low	Extreme

deviations. Every single project, within each organisation, varied through the full range of risk rating options: from low to extreme. This highlights the inconsistent nature of the risk assessments. Some organisations, such as Organisation 3, were more risk averse (e.g. ranging from medium to extreme within one standard deviation in Project 3), whereas others were less risk averse in other projects (e.g. Organisation 4, Project 1). However, despite this, their ranges still did not change significantly when considering all of the responses.

The Risk Assessment Process Is Not Objective

Considering the range and wide distribution of risk assessment scores within this study, one cannot explicitly state that the risk matrix assessment process is objective. The risk rating is, thus, dependent on the person who undertakes the assessment, despite each risk assessor being provided with identical information and using the same organisational risk assessment process.

This finding provides a pathway to understanding decision-making within the water sector, and its role within risk assessment processes. The research implication is particularly intriguing as risk assessments form a key component in determining funding allocations for projects. A high-risk option may not be allocated funding over a low-risk scenario. Therefore, the process by which the risk assessor determines these ratings is influential in the allocation of funds. In many cases around the world, funds for water-based projects are sourced from taxpayers, and therefore, some level of scrutiny into how these funds are allocated is fair and reasonable in the public domain.

Understanding the role of risk assessments in a water project, and particularly its subjective nature, can provide a pathway to implementing new measurement approaches that create a less biased outcome. The public rely on government experts to undertake a reliable risk analysis, for safety, sustainability and planning, among other reasons. However, this study confirms what has been criticised previously (Rae & Alexander, 2017) – that experts differ drastically when asked to quantify risks using their own organisational processes. This prompts questions of 'who is right' and also 'who decides on risk?'

The problem of Public Backlash and Reputation

Each of the participants wrote down their most prominent risks for each project, which was qualitatively coded (see Anna Kosovac, Hurlimann, and Davidson (2017) for the methodology outline). Compared with the quantitative data, this provided a richer source of understanding water practitioner concerns. Figure 5.3 shows the types of risks that assessors saw as most prominent for each project type.

While the fact that the majority of the respondents had technical backgrounds (either in engineering or in operations), two of the three main risks highlighted were not technical in nature, but social. The highest risk noted is the perception that community backlash could alter or stop projects from proceeding. This was taken to refer to a negative public perception that ultimately led to some form of protest or action. The fact that the projects were incredibly varied in scope, from social projects through to implementation of new forms of technology, highlighted that this concern was consistent across a wide range of themes. This is not surprising in a participatory democracy, but it is surprising considering the technical background of the practitioners. A focus towards active public decision-making has become a larger feature

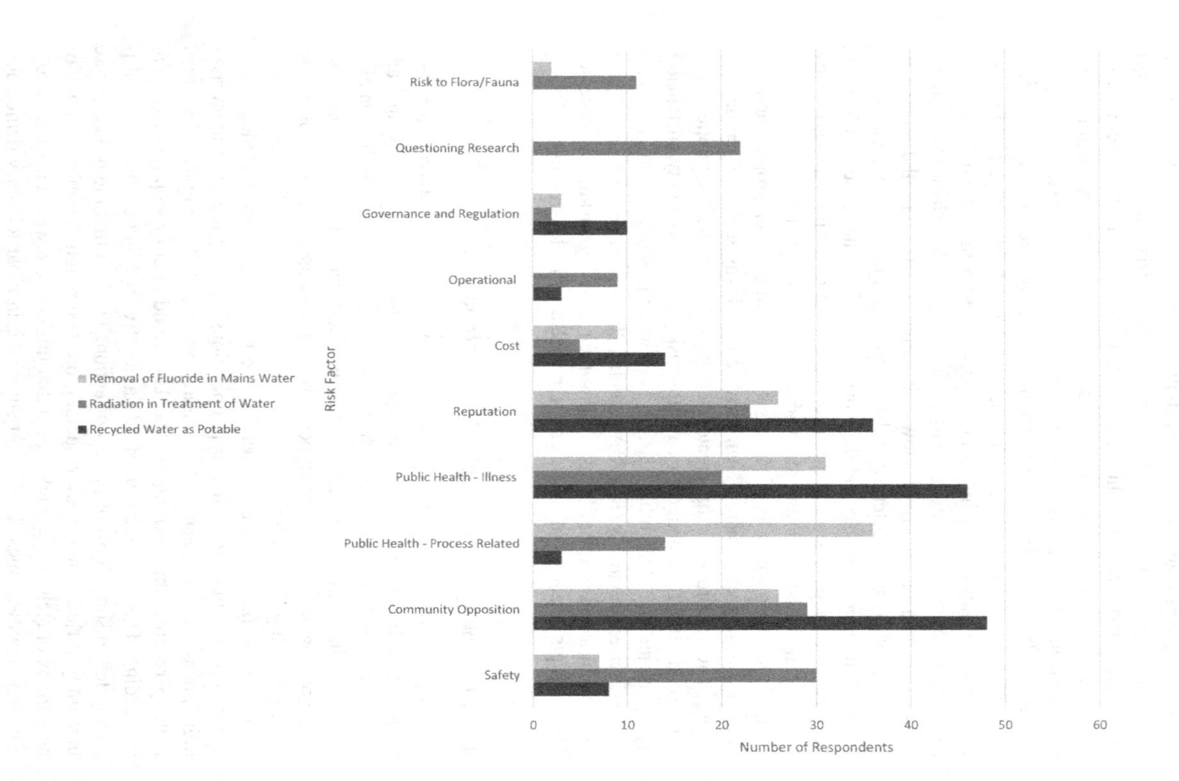

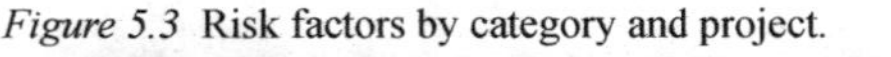

Figure 5.3 Risk factors by category and project.

in the way water projects are managed (particularly in Australia); therefore, community opposition is shown to be front of mind in managing project risks. This was raised for all of the three projects, but had most responses for the potable recycled water project scenario. This is not an unfounded concern, as studies have shown that public perceptions relating to drinking recycled water are generally negative (Dolnicar & Hurlimann, 2009).

Similarly, the reputation of the organisation, also consistently carried a higher position on the list of concerns of water practitioners in the study. This was over and above concerns regarding public health issues. A government institution relies on public trust to be able to function effectively, and to ensure compliance measures for public health are carried through. It may also affect elections in future (see Chapter 6 for a full discussion on this). Distrust may thus affect the organisation's ability to serve the public interest. Reputational risk can also affect government workers directly as negative publicity can serve to impact careers. These water practitioners were acutely aware of the public scrutiny that they would be under if controversies related to projects were to arise, and the subsequent risk to the ability of the organisation to implement policies successfully.

Many respondents mentioned the issue of pro-fluoride groups protesting, while others highlighted that they will be seen as '*giving luddites victory*' (Male Project Manager, 36–45 years old), and heeding to a movement not backed by evidence:

> [I]t's not based on science, it's based on the opinion of people (generally) that have a poor understanding of science. These decisions need to be backed by science. It's a poor response to the problem and focuses on pleasing minorities at the expense of legitimate health benefits for everyone else. Why not allow minorities to use household-based filters to remove the fluoride?
>
> (Male, 18–25 years old)

These comments reflect a certain amount of disdain for segments of the public that do not agree with expertise. It also carries a certain level of superiority, or elitism in its message.

As discussed in Chapter 3, the reputational issue could be as a result of an increasing public distrust in expert risk ratings. This points to the fact that water professionals seem to be cognisant of the effect of community action on projects, and its potential to become a politicised issue in the media. This also highlights the varied role of engineers and operational staff in considering public sentiment and social relations in their historically technical roles.

Dread and Fear of the Unknown

Recognising the difference in the risk scores between people, and their reported qualitative risks, the next place to turn was to the see whether other theories can help us determine what is causing this variance. In this section, we will seek to determine whether there is a psychological risk attribute that is linked to higher risk scores (see Chapter 3 for the full list of psychometric factors). This could result in the one overarching psychological risk attribute or a combination of risk attributes that are linked to higher or lower risk scores, providing an explanatory and predictive tool in the understanding of risk perceptions. Thus, the aim is to determine to what extent a particular psychological affiliation produces a higher risk score in the water industry.

Two key variables explained over 50% of the variation in the risk scores (Figures 5.4 and 5.5): dread related to perceived fatal risk and fear of the unknown. Perceived dread due to fatal risk considers the effect of the risk assessor imagining a scenario within the project that results in a fatality, and in doing so conjures up feelings of dread. Unsurprisingly, with this feeling also comes a heightened sense of perceived risk, and higher risk scores. However, this only had a significant effect when tested on Project 6: the use of radiation in the treatment of water. It should be noted that this result needs to be interpreted in the context of the radiation theme of the tested project, which is notably different to the intent of Projects 5 (Potable to Recycled Water) and 7 (Removing Fluoride from Water). This difference has been noted by a number of studies in this area, highlighting that risk perceptions of radiation and nuclear-technology are high compared to other activities such as driving a car, or smoking, which tend to carry far higher mortality rates (Shrader-Frechette, 1991).

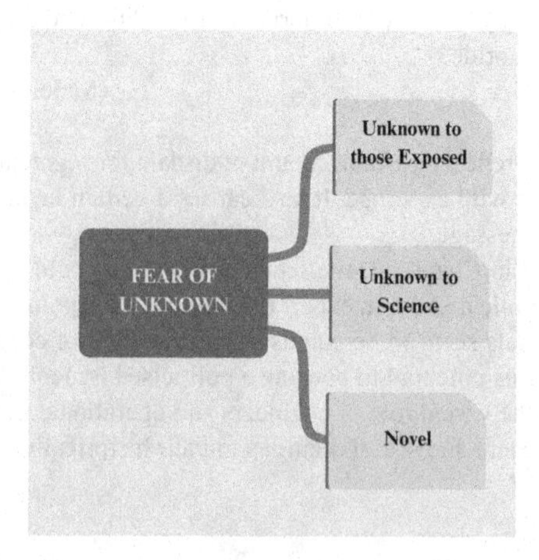

Figure 5.4 Components of 'Dread Related to Perceived Fatal Risk'.

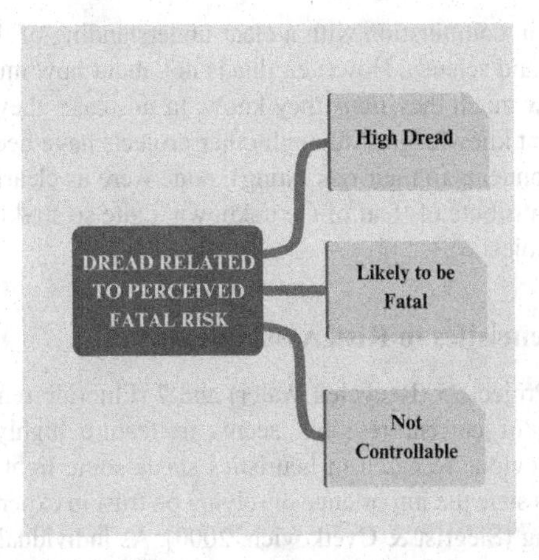

Figure 5.5 Components of 'Fear of the Unknown'.

What are the implications of a feeling of dread on scoring of future projects? It can be concluded from the findings of this research that the more the individual visualises a scenario that evokes a feeling of dread, the higher the risk score. One of the two inputs in determining a risk score or rating is the 'consequence' of the project. The determination of 'consequence' heavily relies on the risk assessor imagining scenarios on which to base their rating. The more dire the perceived imagery, the worse the consequence, thus the higher feeling of dread attached to it. In addition to this, the lower the imagined consequence, the less likely that one would experience a high feeling of dread. In this way, it can be posited that the internal imaginings of the risk assessor, in particular whether they believe a project will result in a catastrophic outcome, point to whether project attracts a high-risk rating. I nevertheless posit that as dread is 'imagined' in these contexts, water professionals carry through their imagined fears into the assessment of the risk.

'Fear of the Unknown' was the other factor that swayed project risk scoring. This highlights that a seeming lack of information or knowledge is a key feature for the higher risk scores. This aspect is not captured within typical matrix risk assessments explicitly, and presents insight into what is not said within these project assessments. Even though the Recycled Water and the Fluoride projects have not been undertaken before, the water professionals would have a clearer understanding of what these are, and the technologies behind them. The radiation treatment project instead presents an unknown quantity so is unsurprising that this in itself raises risk scores. This is suggested by the exhibited influence that 'fear of the unknown' (Figure 5.5), a knowledge-based factor, has on the risk assessor's score. 'Fear of the unknown' is made up of knowledge-based attributes such as the 'newness' and 'unfamiliarity' of the project, in combination with a clear understanding of the risk by both those exposed and science. However, this is not about how much an assessor knows, but how much they *think* they know. In this case, they feel as if they have insufficient knowledge. Although other projects have been rated as 'extreme' by respondents in their risk ratings, none were as clearly linked to the psychological attribute of 'fear of the unknown' quite so starkly as the Radiation project (Project 6).

Trust and Heuristics in Risk Assessments

Compared to Projects 5 (Recycled Water) and 7 (Fluoride removal), distrust or questioning of current research seems to feature highly in Project 6 (Radiation). Previous research in heuristics sheds some light on these findings. Heuristics state the importance of relying on trust in expert knowledge in decision-making (Siegrist & Cvetkovich, 2000). As individuals must rely on some level of knowledge from those around them – recognising that no single person can know everything – trust forms an important element in the sifting of information, especially in current times of factual divergence and disinformation (Kosovac, 2022). Therefore, when new technologies arise, as in the case of innovations such as what is proposed in Project 6, the importance of trust in the researchers that have undertaken the studies is pertinent in guiding the risk perception of water practitioners in their risk assessments (Siegrist and Cvetkovich, 2000). With a lack of information, respondents resorted to heuristics and preconceived notions of the technology (e.g. representativeness bias) to fill the knowledge gap. Radiation carries many negative connotations in the minds of people who do not often deal with this type of technology (Pahner, 1976). Evoking affect (and availability) heuristics, the devastation from nuclear accidents such as Fukushima still carry vividly in the minds of people, especially in the context of Australia, where there is minimal involvement with nuclear technologies; i.e., only one nuclear reactor currently exists. This cultural aversion to nuclear technology could account for the higher uncertainty, and a higher perceived probability of fatalities mentally evoked, suggesting the availability heuristic in deeming this risk of higher calibre than others.

What Can the Cultural Worldview of Assessors Tell Us About Risk?

An assessor's worldviews cannot be distinguished from their risk assessments, or so cultural anthropologists argue (see Chapter 4). How one thinks society should be structured, its process of decision-making and how it deals with minorities are all factors that influence risk. One can sit in one of four groups: Hierachy, Egalitarianism, Individualism and Fatalism. Kahan has taken Douglas and Wildavsky's scales used in Chapter 4 to create a set of scales that are more readily measurable: Hierarchical-Communitarianism, Hierarchical-Individualism,

Egalitarian-Individualist or Egalitarian-Solidarism. These nevertheless carry similar implications to the original scales which reflect positions on how decisions in society should be made. Figure 5.6 represents the outcomes of a survey (using pre-validated cultural-cognition scales by Kahan (2012)) on the same water professionals in the last study. About 68% of those surveyed answered the worldviews test in a way that reflected a Egalitarianism-Solidarism worldview. There is a smattering within the other quadrants, but they are in the minority. Egalitarianism-Solidarism implies community-wide decision-making for public matters, minimal hierarchies of power and a greater focus on the 'common good' rather than the 'individual'. For example, they would be in favour of reducing inequality while also positioning government as the key organisation responsible for making sure everyone's basic needs are met. Conversely, an Egalitarian Individualist would also prefer equality, but they would like to see it happen through allowing individuals greater control to change their lot in life.

This may not be terribly surprising considering all participants are employed within the public sector. Employees are able to decisions on behalf of the population, but these decisions are often under scrutiny and therefore must be transparent. Decisions or allocations of funding by public servants can be overturned, or have pressure applied, by the elected government Water Minister. In this way, democratic principles are still upheld, a central tenet of an Egalitarian worldview. It is less about experts having a final say on the basis of their hierarchical position, but rather allowing a communitarian approach wherein communities can be involved in public sector decision-making. Therefore, the nature of a public servant is congruent with the Egalitarian worldview.

The Egalitarian-Individualist worldview represented a sizable proportion of the respondents as well – approximately 23% of participants. Although not as large as the membership of the Egalitarian-Solidarism camp, it nevertheless represents a noteworthy proportion. Interestingly, those in the Egalitarian-Individualist quadrant consider that the position of a person should not determine what decision-making power they may have – that all should in essence have relatively equal power. However, where they deviate from Egalitarian-Solidarism is the way that decisions should be made. It is fascinating that these public sector workers should consider that decisions are better if made by the Individual, not by the group, as they work in a structure where decision-making on behalf of many is common.

By and large, however, the sheer majority of people responding landed squarely in the Egalitarianism-Solidarism quadrant. It is unclear whether this is as a result of recruitment and attainment of people of similar worldviews, or whether employees within the public sector in water adhere to the culture and worldview of the organisation (see Douglas (2010) for more on the cultural worldviews within organisations). The difficulty here is whether these measured worldviews represent individuals or the cultural make-up of the organisation. This could be a factor reflected in the way that the scales were answered, in

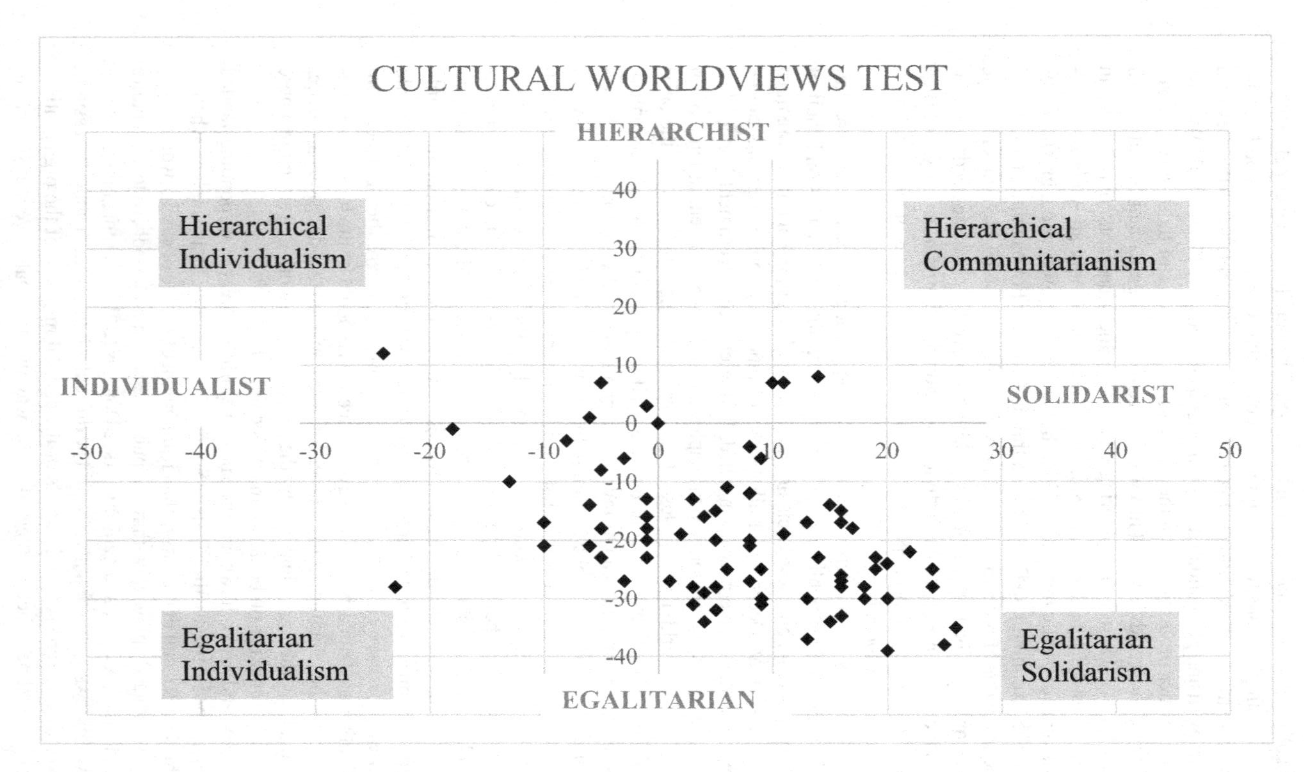

Figure 5.6 Plot of the cultural worldviews test. Each marker represents one respondent.

that the social norms that govern the organisational 'way' are those that fit within one of the quadrants, namely, Egalitarian Solidarism. Rayner (1992, pp. 107–108) states that people within the organisation will "flit like butterflies from context to context, chanting the nature of their arguments as they do so". Furthermore, the longer time one spends at an organisation will also impact the strength of the organisational norms that are impressed upon them, that are in turn performed in their daily work lives. So how does this inform their risk understandings?

To start with, it is expected that those who hold diverse worldviews would generally disagree on risk acceptability (Kahan, 2012). However, those that are more Egalitarian and Solidarist in their views are more likely to care about the environment, to be concerned about risk of climate change, and risk to communities (Kahan, Braman, Gastil, Slovic, & Mertz, 2007). Whereas the more hierarchical or individualist a person's (or group's) worldviews, the less likely they would care about these factors. The issue of blame is often called upon in notions of risk, (see, for example, Douglas's (1992) work on risk and blame). Those that adhere to hierarchy are more likely to reject claims of climate change, while those that were Egalitarian were more likely to find it a dire environmental risk (Kahan, Jenkins-Smith, & Braman, 2011). A recent study highlighted a strong link between those that adhered to environmentalism-solidarist and agreement that anthropogenic global warming is happening. The fact that the majority of water utility practitioners in this survey take on Egalitarian views speaks volumes about the cultures within the organisation, particularly regarding environmental concerns. This is important in considering the nature of the work they do. Environmentalism and water *should* go hand-in-hand but often do not. In particular, this exists in instances of trade-offs or conflict between traditional engineering approaches vs nature-based solutions, or the tensions present in the allocation of scarce water supplies for economic gain rather than to the environment. An Egalitarian would recognise the importance of a community-minded approach that is less about the individual and more about the macro-level effects. As their underlying cultural disposition tends to be more Egalitarian-Solidarist, this also affects the way in which they engage with new information. For example, someone that has a strong pull towards mitigating climate change will likely regulate the 'facts' they accept. As previously noted, an increasing issue in our current society is misinformation and disinformation. This leads many to conclude that the tension is not regarding values, but rather what facts to accept (Kahan et al., 2011). In addition to this, Cohen (2012) argues that identity-protective motivations keep people within their segment of the cultural-cognition scales. This means that someone who is Egalitarian-Solidarist, and everyone around them is as well, they will work harder to retain this position within this segment to reaffirm their identity within the organisation. This, in turn, influences notions of risk and uncertainty – and as a result, decision-making in projects, policies and funding allocations.

References

Cohen, G. L. (2012). Identity, belief, and bias. In J. Hanson & J. Jost (Eds.), *Ideology, psychology, and law* (pp. 385–403). https://doi.org/10.1093/acprof:oso/9780199737512.003.0015

Dolnicar, S., & Hurlimann, A. (2009). Drinking water from alternative water sources: Differences in beliefs, social norms and factors of perceived behavioural control across eight Australian locations. *Water Science and Technology, 60*(6), 1433–1444. https://doi.org/10.2166/wst.2009.325

Douglas, M. (1992). *Risk and blame: Essays in cultural theory*. London: Routledge.

Douglas, M. (2010). *Risk acceptability according to the social sciences* (1st paperback ed.). London: Routledge.

Franklin, J. (2015). *The science of conjecture: Evidence and probability before Pascal*. Baltimore, MD: Johns Hopkins University Press.

Kahan, D. M. (2012). Cultural cognition as a conception of the cultural theory of risk. In S. Roeser (Ed.), *Handbook of risk theory* (pp. 725–759). Netherlands: Springer.

Kahan, D. M., Braman, D., Gastil, J., Slovic, P., & Mertz, C. (2007). Culture and identity-protective cognition: Explaining the white-male effect in risk perception. *Journal of Empirical Legal Studies, 4*(3), 465–505.

Kahan, D. M., Jenkins-Smith, H., & Braman, D. (2011). Cultural cognition of scientific consensus. *Journal of Risk Research, 14*(2), 147–174. https://doi.org/10.1080/13669877.2010.511246

Kosovac, A. (2022). Factual divergence and risk perceptions: Are experts and laypeople at war? *Exchanges Interdisciplinary Journal, 9*(2), 35–55. https://doi.org/10.31273/eirj.v9i2.660

Kosovac, A., Hurlimann, A., & Davidson, B. (2017). Water experts' perception of risk for new and unfamiliar water projects. *Water (20734441), 9*(12), 1.

Neiva, A. (2023). Pascal's Wager and decision-making with imprecise probabilities. *Philosophia, 51*(3), 1479–1508. https://doi.org/10.1007/s11406-022-00586-w

Pahner, P. D. (1976). *A psychological perspective of the nuclear energy controversy*. International Institute for Applied Systems Analysis, International Atomic Energy Agency, Vienna, Austria. https://pure.iiasa.ac.at/id/eprint/617/

Pascal, B. (1958 (1670)). *Pascal's Pensées*. New York: E. P. Dutton & Co., Inc.

Rae, A., & Alexander, R. (2017). Forecasts or fortune-telling: When are expert judgements of safety risk valid? *Safety Science, 99*, 156–165. https://doi.org/10.1016/j.ssci.2017.02.018

Rayner, S. (1992). Cultural theory and risk analysis. In S. Krimsky & D. Golding (Eds.), *Social theories of risk* (pp. 83–116). Westport, CT: Praeger Publishers.

Shrader-Frechette, K. S. (1991). *Risk and rationality: Philosophical foundations for populist reforms*: Berkeley: University of California Press.

Siegrist, M., & Cvetkovich, G. (2000). Perception of hazards: The role of social trust and knowledge. *Risk Analysis, 20*(5), 713–719. https://doi.org/10.1111/0272-4332.205064

Simmons, M. B. (1995). 174 God and the Gods. In M. B. Simmons (Ed.), *Arnobius of Sicca: Religious conflict and competition in the Age of Diocletian*. New York: Oxford University Press. https://doi.org/10.1093/acprof:oso/9780198149132.003.0006

6 Implications

Entering a place of work, one cannot help but bring in their own values, beliefs and backgrounds into their daily tasks. This often forms the backbone to what we do, and is part of the reason that many align themselves (when possible) with a workplace that does not inflict moral injury (i.e. discordant values to one's own). However, an individual's risk perceptions purely reliant on one's values and experiences cannot fully explain risks measured in the data described in Chapter 5. The culture and norms within an organisation feed into such assessments in ways that are often insidious and gradual. This has also been shown to change with seniority within an organisation. This does not refer solely to how high up the hierarchy one is, but rather in combination with how long one has spent at the organisation. The higher the seniority in the organisation, the more likely that they will have stronger dedication to their employer (Sewell, 1971). As such, institutional effects will play a role within how risk is perceived for an individual. The argument that will be developed here is on the basis of reputational risk, dread and how we can explain these results within this context.

It is increasingly becoming clear that current risk processes, especially in areas that overlap with social and political factors, cannot be deemed as objective. Studies have addressed this gap, recognising the clear role of the social sciences in addressing this space (see, for example, Douglas (1992); Fischhoff (1995); Kosovac, Davidson, and Malano (2019); Lupton (2013); Pidgeon, Slovic and Kasperson 2003; Slovic (1992); Zinn (2008b)). Individual psychology, affect, cultural factors and our own environment significantly alter the way we score risks. This exists despite descriptive organisational processes in detailing how to score consequences and probabilities in assessments. For many projects, especially those assessed at a concept stage, which is often the case when undergoing options analysis, the risk assessor must essentially 'crystal ball' consequences to determine what they *think* may occur in the project. For example, one could imagine a number of processes failing, resulting in a catastrophic event, while this mental episode might not even occur to another assessor. Therefore, each person will assess a project based on what they believe will happen. This is likely to be the reason for the

DOI: 10.4324/9781003432647-6

variation in the scores. This is also underscored by the qualitative findings, with each assessor coming up with differing risk scenarios, and even when they voice similar risks, the consequence of each risk alters again between each individual. To compound this issue further, the likelihood score is again a matter of personal prediction. For some risks, such as a pipe failure, or a storm event, assessors can be provided with adequate data to form a consequence and probability score. For example, a water authority may have data on diameter 300 mm PE water mains, and when each has failed. They can use this data to predict the life of a typical 300 mm diameter PE main, and therefore the likelihood can be mathematically calculated. Similarly, if they have data on failure modes, and effects for each of the failed pipes, they are able to determine the likely consequence for their new, proposed pipe. However, the failure of the pipes is often not what is scored the highest in risk assessments, as shown in Chapter 5. The high scores relate to political and social elements that cannot be easily quantified away like a pipe failure.

An important finding from this study indicates that there may be an issue with the current risk assessment approach, in that bias is introduced into the scoring. This could prompt a reassessment of how risk is calculated across the organisation, in particular to ensure a better fit with the organisational risk appetite. This may also prompt the need for more training into current risk approaches, or other methods with which to ensure more consistent scoring practices. IEC 31010 (International Organization for Standardization, 2009a) suggests that more than one expert undertake the risk assessment to ensure that there is some level of consistency. Hemming, Burgman, Hanea, McBride and Wintle (2018) assert that there should be a minimum of four experts consulted as part of the risk assessment process. However, this approach can introduce many other biases into the process. Heuristics highlight the importance of trust as an element in decision-making in uncertain times (Tversky & Kahneman, 1973). Relying on the knowledge of experts to provide advice in certain areas may be needed, and with this reliance comes trust. Subsequently, when making the choice of which experts to consult, the practitioner may turn to those they trust, or even those that they share an affiliation or similar views with. This should be further explored in future research to determine whether this choice of expert in itself promotes the bias that is to be minimised, especially in the face of experts choosing other experts that may share views with.

However, this may only correspond to one of the risks. Other risks, such as financial, regulatory, but especially community-based risks, can be harder to predict. The main reason is the sheer number of factors affecting how a social-based scenario comes about. It could be the socio-demographics of a region, the political slant, employment status, education of those around, other projects that might come into conflict, past experience of the community, etc. Therefore, when making an assessment, risk assessors may use some level of heuristics – in particular, availability heuristics (refer Chapter 3) can be as a result of previous experience, or perhaps information they deem salient.

With Increasing Technological Changes, Risks Are Getting Riskier

Ulrich Beck's (2006) theory on increasing risks felt by today's society is not just an abstract thought to be considered and tossed away. Rather, it affords us reflection on how we have shifted as decision-makers, as public sector providers, in our conceptualisations of risk. Have we been more obsessed with risks over time, how has this changed our practice?

In the environmental sector, and notably the water field, situations that we have not previously experienced before are now coming to light. These trends include:

Increasing difficulty in prediction.

Weather patterns and climate models have become increasingly difficult to predict. Past climate and water supply data may no longer be valid in an uncertain future (Wasko, Sharma, & Johnson, 2015). Dealing with uncertainty (and wrong predictions) is taking up more and more of a water professional's time. Our ability to deal with uncertainty, however, differs greatly between individuals. But there are some common factors that appear sector-wide. Namely, the water sector is heavily reliant on the modelling and expertise that has its foundations in the engineering and natural sciences. This training does not often result in sitting comfortably in 'not knowing' but is predicated on a reliance on quantitative measurement which is underlined by notions of certainty. I (together with colleagues Kris Hartley and Glen Keucker) have reflected on the dangers of this type of technocracy, and overreliance on figures to inform (Kosovac, Hartley, & Kuecker, 2021). The danger exists in the seeming illusion of safety in figures. This is not to presuppose that the attempt to measure uncertainty is callous or altogether unhelpful. Quantitative approaches have their merits for many aspects of water planning, such as demand tracking, pipe upgrades, supply mapping, all of which should always be up front in their assumptions, especially those related to forecasting.

Where these type of assessments become undone is in situations of high consequence and extremely low probability. The realm of these risks exists in a region that is often ignored due to its perceived unlikeliness. These 'black swan' events are arguably becoming more prominent within the landscape of areas such as the climate – once again a call-back to Beck's Risk Society. These risks can be catastrophic, often incalculable, but most importantly, uninsurable. "Black Swan Logic makes *what you don't know* far more relevant than what you do know. Black Swan events can be caused and exacerbated by *their being unexpected*" (Taleb, 2008, p. 2). Nevertheless, Aven (2014) suggests a number of actions that can be taken when preparing for Black Swan events, namely, focusing on signals and warnings. When working within an industry for a while one often develops a 'feel' for how things are tracking. Running, for example, is a circumstance in which we are aware of how our body feels. We can begin to feel when our body is becoming weary, and at risk of collapse. It may be at this point that we decide to stop running. In this way, having an understanding and some level of intuition in water planning can help to track these warnings and signals to avoid a system collapse. To be able to do this effectively, it is imperative to have a collective mindfulness as an organisation to pick up on small shifts and changes to manage such risks (Aven, 2017).

As water decision-makers, there is nigh a single practitioner that would not recognise the importance of their role in providing a life-sustaining uninterruptable resource. The public trust water practitioners to keep this supply running, often with minimal regard for the complexities of insuring such a supply. The logistics of water delivery are hidden, out of sight and represent that mostly 'invisible city' of pipes hidden underground that is only noticed when something goes wrong.

Water must be ensured, and insured

The social contract between water providers and the public is increasingly getting harder to meet over time. Risks from climate change in particular present global risks that can be difficult to measure. Not only is this an issue of a decline in freshwater resources, but the increasing frequency of flooding and fire events also have substantial detrimental effects on the safety of the supply (Wasko, Nathan, Stein, & O'Shea, 2021). Furthermore, many cities around the world are experiencing heightened levels of urbanisation – a phenomenon not merely about an increase in urban populations, but rather a shift in the proportion of people living in urban areas in comparison to rural ones.

Contention in the allocation of water resources between various water users (urban dwellers, environment, cultural uses) means that water decision-makers need to have a wider range of knowledge to assess a burgeoning range of risks that are no longer technical in scope. For example, declining fauna populations, risks to social cohesion (as discussed in Chapter 1, the impact from water restrictions), class, age, resource availability, cost and processes of decision-making all factor into risk imaginings. Costs cannot be separated from the supply of water in capitalistic systems. This has resulted in what Sofoulis (2005) dubs as the increasing 'customerisation' of water users, that water users are seen less as citizens or users but rather enacting a direct relationship between a water user and money. When issues of inflation come to the fore, this increases the impact that water services play on those that use them, and this disproportionately affects lower socio-economic strata. All of these factors highlight the varied aspects that water providers need to be aware of, from cost-of-living stresses, through to equity and also environmental impacts.

As new technologies come to the fore, we must remain vigilant

Chapter 5 highlights a study at the forefront of the constant battle between risk and innovation (Kosovac and Davidson, 2020). Within the study, I picked up on relentless risk aversion plaguing the water sector. And I use the word 'plaguing' consciously, as at the time of writing up the study, I exhibited a sense of admonishment towards the trend. In light of the ever-changing nature of risks, and the rate of technological change, I cannot say that I hold the same view. It is ridiculous to say that risk aversion does not stifle innovation, but risk aversion can also be considered a form of the 'precautionary principle'. The Precautionary Principles state that new technologies should be viewed as risky unless evidence arises otherwise (Sunstein, 2005). It presupposes that we take action assuming danger, where there might not be any. An example of this is the European Union (EU) insecticides case. When the effective neonicotinoid insecticide was released for use in Agriculture, it provided a boon for farmers worldwide who needed to insure their crops from insect damage. However, the impact of neonicotinoids at time of release was unclear, the science not yet definitive. The EU first approved these insecticides in 2005, and partially banned them in 2013, citing the precautionary principle (European Commission, n.d.). There were reports of harmful effects on bees, and other invertebrates which often lead to death. Although this was not yet scientifically proven, the EU imposed a ban on three neonicotinoids until it waited for more conclusive evidence from the applicants on the safety of its use. The information received for the re-evaluation in 2018 confirmed that the previously identified risks were still valid, i.e. that there was a link between neonicotinoid use and honeybee decline, forming the basis for proposals to fully ban these three substances in outdoor areas. This represents an example of the use of the 'better safe than sorry' principle in utilising precaution in situations where the science is not conclusive.

In the case of the water sector, the invocation of the precautionary principle could relate to examples such as new industrial waste that enters treatment plants. It is unclear whether treatment plants can remove any potential 'new' toxins, and there is little scientific certainty surrounding the impact on sea life when the treated water is discharged into the ocean. The precautionary principle would say that there should be increasing monitoring of the water quality, a ban on the dumping of the industrial waste down the wastewater network until it can be scientifically proven that it presents little risk to the health of the community and wildlife. This reflects a proactive stance to protect public health and the environment in the face of scientific uncertainty. With the fast-paced nature of technological changes, the vigilance required for water authorities must be increased, and new technology should be carefully scrutinised, applying the precautionary principle where relevant. As such, the risk aversion exhibited in water industry practice is appropriate to ensure the ongoing safety of community members.

Our Feelings and Risk Assessment

The dread we feel when facing risk choices has been the largest predictor for increased risk perceptions. Dread represents that sinking feeling of impending doom that is evoked when considering risk imaginings. In other words, those that catastrophise, thus bringing about a sense of dread, are likely to rate risks higher than the average risk assessor (see Chapter 5). It is then up to the determination of the organisation to define its risk appetite and thus allocate projects accordingly. A more risk averse organisation may opt to allocate a project to a project manager with more life experience, as this could be reflected in their risk approach; i.e., research has shown that with increasing age comes more risk averse behaviours. Whereas an organisation that may strive to encourage new and innovative projects could look to allocate these to those less experienced persons. Alternatively, an option could be to use these results to create a balance, creating a diverse risk assessment panel that varies in age. The IDEA framework (discussed later in this chapter) is a useful process for structuring these types of risk panels.

Chapter 5 had reported on the finding that reputational risk is a key concern among public water sector officials in Melbourne. Trepidation related to this overrode other risks such as public health, safety and cost. Democratic principles allow citizens to select their representatives. Elected representatives are exactly that: people placed in a position to embody the general views of its citizens. In measuring the qualitative responses to prevalent risks for each project, 'community opposition' was a risk factor that featured the most. This was closely followed by risks related to reputation of the organisation. Both of these topics were ultimately related to a negative public perception of the project that led to some form of protest or action. This is not surprising in a participatory democracy, with the State Government's position as water authorities' key shareholder. Naturally, the decisions of democratically elected politicians will be influenced by a keen regard for community support, or lack thereof (Kosovac, Hurlimann, & Davidson, 2017). However, it does not necessarily follow that the views of politicians will directly mirror broad public opinion. Sjoberg and Sjoberg-Drottz (2008) have explored the difference between politician and layperson perspectives of nuclear waste issues in Sweden, noting that although political leaders felt they had a differing viewpoint from the general community, their measured views and perceptions were notably similar. This suggests there may be a disconnection between what politicians *think* the community wants compared with what they actually want.

In a recent study by Sivagurunathan, Kosovac and Khan (2022), they explored water practitioner perceptions of potable recycled water in Sydney, Australia and compared this to local politicians' view. Figure 6.1 highlights the difference in expectations in developing a potable recycled water project in Sydney within the next 30 years. Politicians were generally against the scheme, with a mean at the mid-point, while water practitioners were much more optimistic about the likelihood of such a scheme developing in the future. The politicians continually cited the lack of community approval as a large impediment to the recycled water, despite evidence saying otherwise.

A focus towards active public decision-making has become more ubiquitous in the way water projects are managed in Australia, and therefore unsurprisingly, community opposition is shown to be a key concern in managing project risks. Public sentiment, or backlash, can halt or stop projects altogether, and thus plays a key role in the project management process. Hurlimann and Dolnicar (2010) analysed the reasons why an indirect potable recycled water system proposed in 2006 for Toowoomba, Australia failed. A plebiscite was held, and over 60% of the community voted against the scheme.

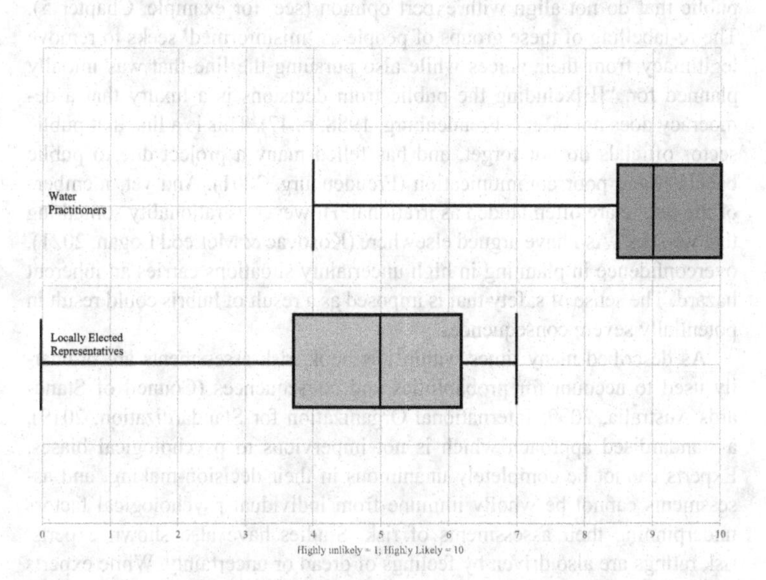

Figure 6.1 Industry versus politicians ranking (1–10, from highly unlikely to highly likely) of developing a potable reuse project in Sydney in the next 30 years (n = 20).

Source: Adapted from Sivagurunathan et al. (2021).

The authors found that the failure of the plan was not just associated with community opposition, but also due to political involvement, the timing of the plebiscite, vested interests and information manipulation. Public opposition also had a role to play in other potable reuse schemes including that proposed for San Diego in the United States in the 1990s (Mills, Karajeh, & Hultquist, 2004), and for a Desalination Plant in Sydney in the 2000s (Davies, 2006). Po and Nancarrow (2004) have attributed a DAD 'decide, announce, defend' approach to the demise of these schemes (Forester, 1999).

As Assessors, Can We Retain Rationality?

Many from the natural sciences and engineering may argue that there are elements of decision-making where rationality must be maintained. When the public sector is increasingly scrutinised over its decisions and its spend, the idea of a 'rational' individual at the helm of public decision-making is often seen as desirable. This is often called upon when considering the role of communities in such debates. Laypeople are seen as offering unique insights into projects and issues that may have not been considered by a project manager. These interactions are at best glorified counselling sessions, and at worst, a form of placation. There is often contempt, or disdain for segments of the public that do not align with expert opinion (see, for example, Chapter 5). The re-labelling of these groups of people as 'misinformed' seeks to remove legitimacy from their voices while also pursuing the line that was initially planned for. "[E]xcluding the public from decisions is a luxury that a democracy does not offer" (Freudenburg, 1988, p. 47). This is a line that public sector officials do not forget, and has felled many a project due to public backlash and poor communication (Freudenburg, 2001). And yet, members of the public are often lauded as irrational. However, is rationality something that we seek? As I have argued elsewhere (Kosovac & McLeod Logan, 2021), overconfidence in planning in high uncertainty situations carries an inherent hazard. The sense of safety that is imposed as a result of hubris could result in potentially severe consequences.

As described many times within this book, risk assessments are ordinarily used to account for probabilities and consequences (Council of Standards Australia, 2009; International Organization for Standardization, 2019), a standardised approach which is not impervious to psychological biases. Experts cannot be completely unanimous in their decision-making, and assessments cannot be wholly immune from individual psychological factors underpinning their assessments of risk. Studies have also shown experts' risk ratings are also driven by feelings of dread or uncertainty. While experts can be swayed by heuristics and other psychological factors, studies have determined that experts are more homogenous in their assessments of risk in comparison to assessments undertaken by the general public (Drottz-Sjoberg, 1991; Margolis, 1996). This was further confirmed in a study by Ochi (2021), stating that scientifically trained experts are less vulnerable to be swayed by cognitive and social forces due to their consistent and habitual sourcing of information. Despite these findings, there has nevertheless been an ongoing debate in the last 30 years on the premise of whether experts are the objective, rational decision-makers that many claim them to be. Conversely, do we want purely rational decision-makers (assuming this is possible)? A study on chess players showed that they performed better when relying on heuristics than when they purely rely on risk as analysis (Slovic, Finucane, Peters, & MacGregor, 2004). Similar findings have been reported on those conducting security screenings at airports (Slovic et al., 2004). This premise is also well encapsulated in a study by Braman, Kahan, Slovic and Gastil (2006): "[l]ike members of the general public, experts are inclined to form attitudes towards risk that best express their cultural visions". The only difference, they argue, is that experts are more likely to use their technical knowledge and rationality in this judgement. It is this knowledge that will also factor into their risk assessments, ensuring experts remain a necessity in decision-making.

How to Conduct Effective Expert Risk Elicitation

We have reached the point where we recognise the fallibility of traditional approaches to risk and how they can be affected by a myriad of personal and cultural issues. As water professionals, our aim is not to overtly bias our decision-making, but to be reflective of our own psychological and cultural makeup. This is a dire need in the public sector to ensure that we are making transparent decisions on behalf of many.

No matter what ontological line you take, whether this be that you see risk as real or whether (like a true constructivist) see risk as something that is socially created, there is still a practical component in that we are still asked to measure and quantify risks regardless. Whether we can ever truly 'know' the risk is another divergence in risk theory. Psychological proponents say that one can never truly know it because our minds will constantly skew what is before us, while sociological theorists would say that we also cannot know it as a society because it does not exist out there, waiting to be found. Natural sciences posit that yes, the risk both exists and we can know it, if we know what we're looking for. Regardless of the position one takes, a pragmatist recognises the need, especially in our current society, that risk needs to be reported on, recognised and acted upon to ensure some level of insurance for the general population. Right or wrong, this is the expectation, and responsibility, placed on public sector workers.

So, recognising that our cultural and psychological effects will skew our risk imaginings, what is a way that we are able to ensure a structured process to elicit expert opinion?

A structured elicitation protocol called IDEA (Investigate, Discuss, Estimate, Aggregate) is a useful framework for such a feat. This tool has particularly been useful in conservation science, and environmental decision-making, especially in times of uncertainty (see, for example, Adams-Hosking et al. (2016)). It allows for the structured decision-making of experts to be able to minimise the mistakes due to the range of biases that people are exposed to (as described in Chapters 3 and 4).

Structured protocols create the safe space to be able to sufficiently elicit expert opinion within specified bounds, allowing for critical appraisal, review and consensus.

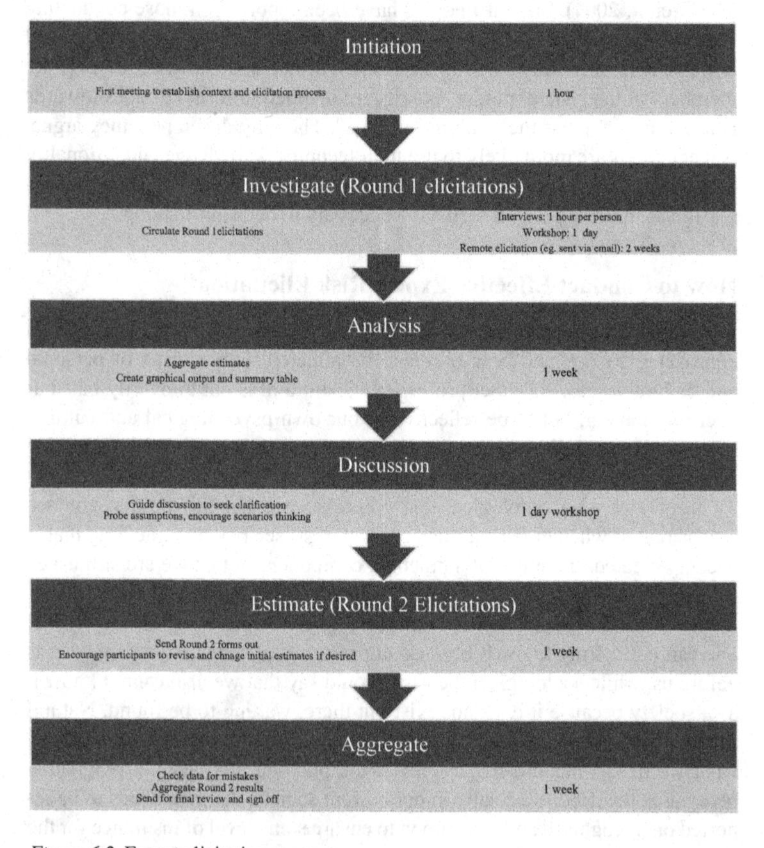

Figure 6.2 Expert elicitation process

Source: Adapted from Hemming et al. (2018).

The first process is to ensure effective preparation and planning for the assessment.[1] Hemming et al. (2018) highlight that this is a key to a successful elicitation, and requires the following steps:

1 **Create a Timeline**

Developing a timeline of the elicitation process that includes what each task involves, the people involved in each stage, as well as when they will be conducted. Recognise that time should be built in to factor in delays in responding, or approval hold-ups. Generally, a single group of experts may take between 2 and 6 weeks to fully complete the elicitation process (Hemming et al., 2018).

2 **Allocate Roles**

Roles that should be assigned in a structured expert elicitation include:

1 Problem Owner – requests the elicitation
2 Project Manager – manages the logistics, timings and people involved in the process
3 Facilitator – in charge of facilitating and guiding each elicitation session
4 Analyst – processes responses and combines estimates.

3 **Develop Questions**

Consider the questions that need to be asked within the process. This could relate to opinions based on likelihoods or consequences relating to certain risk problems. It is generally best practice to ask no more than 20 questions for each session (Burgman, 2016). Basic steps include:

a Develop clear and critically important parameters for the questions.
b Review questions with two independent experts to remove any ambiguities.
c Provide any background information that is relevant to the discussion.
d Questions should be in a format that is familiar to those in the group, e.g. sent via Word Document/pdf, online form or via text in email.

Examples of questions might be:

– How many severe drought events (three years +) will be expected to occur within the next 50 years that will lead to public water restrictions in the area of Metropolitan Melbourne?

- When you think of the reasons that make this likely to happen, how sure do you feel that water restrictions will occur within the next 10 years? 20 years? 30 years? State probability 0–100%. Provide thoughts/ justification for response.
- When you think of the reasons that make this unlikely to happen, how sure do you feel that water restrictions will occur within the next 10 years? 20 years? 30 years? State probability 0–100%. Provide thoughts/ justification for response.
- Consider the balance of evidence from the two sub questions and revise a final probability figure.

- Follow-up question may be: imagine that there is a severe drought event that has lasted over five years in Melbourne. What would be the effect of public water restrictions on communities and wildlife over this time? Consider social, political, fiscal and technical elements related to this question.

4 **Determine the expert panel**

It is good practice to aim for approximately 6–12 people on the panel, and the main basis for selection should be on account of requisite expertise, but most importantly whether the person can understand the questions being asked. Consider the specialisation and skills of those involved, while also ensuring diversity in the cohort. Diversity should consider elements such as gender, age, education, specialisation and culture background. It has been shown that diversity within these elements allows for cognitive-diversity, an important factor in considering risk perceptions. Another key factor that should be considered is power dynamics. Recognise where group members are reporting to one another, hierarchical factors, close friendship, etc. These power dynamics can serve to sway perspectives, providing an unequal level of power or 'voice' in judgements. The facilitator should be aware of such dynamics, and where possible, avoid them. If this is not possible, consider asking feedback from those at the lower power end first, and then going to others afterwards.

5 **Send out Round 1 forms (the list of questions compiled) for each expert to complete individually**

This gives the opportunity for each expert to take time to consider the material presented and formulate their own risk ratings and justification without other expert influence.

6 **Analyse Round 1 results**

Compile Round 1 probabilities and consequences in a graph to show the different ratings from the expert group for each question.

The analyst should collate all responses, and aggregate them into a graph for further discussion.

7 **Plan Discussion workshop**

Create a discussion forum via workshop online or in person to discuss each finding. The facilitator then elicits conversation surrounding the estimates posed. The variations in responses should be discussed, as well as the justifications put forward from each expert. The aim of the workshop is not to develop consensus, but rather to highlight the evidence and justification from others to inform each other's thinking.

8 Send out Round 2 forms to the experts

Send the same questions for each expert to complete individually following the workshop. This allows time for reflection and review following the workshop discussion.

9 Final Analysis

The final assessments are aggregated by the group analyst and used as the final answer to the expert risk elicitation. This forms the basis for the risk assessment and decision-making going forward.

See Figure 6.2 for the full outline of the IDEA process.

Role of the Community in Risk Decisions

In the 1964 movie, Dr Strangelove, General Ripper turns to Captain Mandrake amid approaching gunfire and announces that [fluoride] is introduced into 'our precious bodily fluids' without the 'knowledge of the individual', making spurious links to the Second World War and Communists. The reason that this is called on in a movie like Dr Strangelove is to shine light on the conspiracy thinking in a satirical way, highlighting the preposterousness of the claim (Carstairs, 2015). However, this is not a claim that is so far from reality. We have seen anti-fluoride movements globally, stemming from the 1950s and continuing today. Carstairs (2015) argues that this thinking has been reduced to the fringes of society, but upon examination of the list of fluoride-free water supplies around the world, it is difficult to see it as so.

On 25 October 2010, residents of Waterloo, Canada took to the ballot box to decide whether fluoride should continue to be added to drinking water. Months of campaigning from both sides led to a narrow win for the 'remove fluoride' team. With 50.3% of the vote, the majority was only slight, winning by a mere 195 votes (Canadian Dental Association, 2010). The vote was non-binding, but the City Government nevertheless decided to adhere to the result, removing fluoride dosing from its water network. The scientific consensus by this stage was astoundingly in favour of low levels of fluoride within drinking water and its ability to reduce caries across populations (Centers for Disease Control and Prevention, 1999). Despite the evidence, the additive was voted down. What led to this decision, and how did it get to this point?

These days the anti-fluoride movement is closely connected to conspiratorial thinking, in a vein not dissimilar to General Ripper's claim that fluoridation is 'monstrously conceived' and the most 'dangerous communist plot we have ever had to face', a link to the enemy applying seemingly malevolent tactics to the US population. The nature of trust in authorities nevertheless plays a leading role in the risk perceptions of community members, as is shown by the Waterloo case. It is particularly key when considering that laypeople must gain knowledge from those that are deemed experts to inform their decisions and thinking. This is further exacerbated by the properties of water, the fact that water can be unsafe to drink and our senses are not effective at picking up on this contamination (a callback to Beck's Risk Society in Chapter 4). The water itself may not taste, smell or look, unsafe using our latent senses, requiring that people rely on information from official sources to determine whether to drink the water or not. Fitchen (1990) argues that in Individualistic societies (such as the United States), people tend to place greater confidence in their own senses, rather than that of the authorities. Much of the rhetoric stems from the desire for freedom to appraise their own risk. This aligns with the psychological understanding of voluntarily assumed risk as leading to a lower risk perception. This issue of trust in authorities is one of significant concern within the water practitioner community and can also be linked to the tendency to want to reduce reputational risk, and in turn to reduce the chance of community backlash.

Trust, Risk Perceptions and Expertise

Trust in expertise matters in decision-making regarding risk trade-offs in areas such as new technologies, public health and natural disaster management (Groothuis & Miller, 1997; Siegrist, 2021; Siegrist & Cvetkovich, 2000; Zinn, 2008a). Siegrist, Gutscher and Earle (2005) sought to determine whether high levels of trust towards authorities reduce risk perceptions in individuals. They consider the role of general trust, the extent to which one believes that people can be usually trusted and general competence, which considers how 'under control' things are. General trust and general competence were found to be negatively correlated to risk perception. That is, the more trust one instils on the organisation/person presenting the information about a proposed low-risk action or technology, then the lower the risk perception felt by the person. Trust is crucial where knowledge is lacking, particularly in the face of uncertainty. Considering there is consistently an incomplete scientific knowledge in the general population, the impact of trust on risk perceptions is highly influential. If trust is not present, then the ensuing attempt at conveying risk knowledge is not likely to be received or accepted (Earle & Cvetkovich, 2001).

So, what leads a person to develop trust in a company, government or institution? A crucial finding is that value-laden narratives play a key role in trust and therefore, adoption of appropriate risk framings (Ma, Dixon, & Hmielowski, 2019). In particular, the sharing of salient values in the 'stories' a company/government/organisation articulates, and its level of alignment with your own personal values, tends to significantly influence your trust in them. In considering evidence surrounding trust and risk perceptions, Slovic

(1990) found that of the US public surveyed, most people viewed X-rays and prescription drugs as being low risk with significant benefits. These findings were linked to the high level of trust reported by these participants towards medical practitioners. However, when considering industrial radiation, it was seen as high risk generally by those being surveyed, despite evidence to the contrary provided by experts (Ochi, 2021). This was linked to the low trust in governments and those that manage risks associated with these radiation technologies. Subsequently, those that do have high trust in experts perceived fewer risks and greater benefits associated with a new piece of technology (Siegrist, Cvetkovich, & Roth, 2000). In this way, trust towards expertise plays a direct role in effectiveness of the communication of risk, and the actions taken by individuals to address it.

Attitudes Towards Expertise

Related to issues of trust, there has been a seeming public disregard of science which has heralded perceived decreasing trust in scientists and experts in modern democracies (All European Academies (ALLEA), 2018). A dichotomy of 'facts' vs 'untruths' has been utilised in discourse across many issues, whether this be on the radiation impacts of wireless internet in our home, the effects of wind generation farms on local health or even the safety of nuclear power plants. Despite experts playing a critical role in understanding and communicating risks of new technologies and disaster management, public discussions which repudiate facts and information presented by experts are rife (see, for example, the COVID vaccine public debates (Berman, 2020)). In the absence of trust in experts, people may turn to sources of information that are shared by those they trust and share values with (Siegrist & Cvetkovich, 2000). This, combined with the inaccessibility of academic scientific knowledge, often results in a greater reliance on alternative sources of information available through blogs, YouTube videos and other online mechanisms which may not be evidence-based and may be unverified (Lewandowsky, Ecker, Seifert, Schwarz, & Cook, 2012). Those who resort to these types of platforms to source their daily news consider them fairer and more credible compared to traditional news sources (Johnson & Kaye, 2004).

As a result, there has been increasing pressure placed on social media platforms such as Facebook and Twitter to monitor and reduce misinformation, while promoting 'legitimate' sources in their algorithms, making them more visible to users (Facebook for Business, 2020; Ghosh, 2020). There is little point in purely prioritising information from experts, if it so happens that public distrust exists against the institutions that report them (Ochi, 2021). Prominent sociologist, Anthony Giddens (1991), highlighted this challenge when recognising that "the nature of modern institutions is deeply bound up with the mechanisms of trust in abstract systems, especially trust in expert systems", a view similarly echoed by Slovic (1993). But do these tensions we perceive in public discourse translate to real attitudes towards expertise?

The empirical findings on this topic do not substantiate the perceived conflict and distrust between experts and laypeople. A number of studies highlight that, in fact, there is general public trust towards some experts, particularly medical professionals, engineers and scientists (CONCISE, 2020; Sanz-Menendez & Cruz-Castro, 2019). Furthermore, COVID-19 has brought with it a heroisation of medical professionals and epidemiologists in the public eye, which in the case of Australia is exhibited through mass-produced items such as t-shirts and bedspreads glorifying the Victorian Chief Health Officer, Professor Brett Sutton (Gillespie, 2020). This is a global trend that has featured, for example, White House Chief Medical Advisor Anthony Fauci in the trend 'Man Crush Mondays' (Tillman, 2021) and an 'unofficial Dr Bonnie Henry fan club' in Canada (Woods, 2020). Public health experts are valorised in a fashion that is not indicative of a public wariness towards expertise.

For environmental issues such as climate change, there are mixed opinions towards expertise. Studies report on the high levels of trust from the public towards information from climate scientists (Bickerstaff, Lorenzoni, Pidgeon, Poortinga, & Simmons, 2008; Malka, Krosnick, & Langer, 2009; M. C. Nisbet & Myers, 2007), while other studies find that government general science research is trusted generally, yet climate science is less trusted by the public (Myers et al., 2017).

One area that is overwhelmingly backed by ample evidence is that trust (and credibility) is highly dependent on personal political ideology (Bolsen, Palm, & Kingsland, 2019; Brewer & Ley, 2013; Malka et al., 2009). This subsequently affects the efficacy of message (including risk) communication from a variety of sources, in particular that people seek information from those that are ideologically aligned. Hmielowski et al (2013) found that in their US study, those that align with conservative values and consume conservative media were more likely to have lower trust in science than their non-conservative counterparts. This is a finding that has been further confirmed by other studies in the literature (see for example, E. C. Nisbet, Cooper, and Garrett (2015)) particularly in considering the effect of cognitive dissonance: the rejection of information that is contradictory to current beliefs and values. The impact of the psychological practices in cultural cognition and defence motivation also illustrates a role in the likelihood of information acceptance. To elaborate, information that may challenge the beliefs that underpin one's identity may be less likely to be adopted, and more likely to be subconsciously resisted (Giner-Sorolila & Chaiken, 1997). As this information may pose a threat to one's own self-perception, particular facts may be avoided that clash with self-proclaimed identity (Kahan, Braman, Slovic, Gastil, & Cohen, 2009).

Dissonant information often creates conflicts within the ideological identities of people, which can lead to a negative affect towards the scientists delivering the message. This is most often exhibited in the context of climate change and environmental degradation. Bolsen et al. (2019) found that incorporating climate scientists in a national security message on climate change decreased the respondent's perception of the risk of climate change, contradictory to the aims of delivering the message itself. Despite the effect of dissonant information, if the knowledge transferred is provided by those one is ideologically aligned with, they are more likely to accept the information. For example, a Republican voter is more likely to accept climate information from the Republican Party than from other ideologically non-aligned sources (Bolsen et al., 2019). As such, individuals seek out and accept information that is in line with their own worldviews. Political polarisation around science has the potential to depress trust in science, regardless of where one lies on the ideological spectrum. Therefore, although trust in experts does exist, the impact of ideology on these relationships cannot be ignored.

A Note on Western and Colonial Underpinnings of Risk

I recognise, as writing this book, that the notions of risk that I reflect and report upon are those that have been developed and propagated within the Global North. This is not to presuppose that these are overarching risk understandings that are complete in their scope. Rather, in applying a decolonisation lens through the risk paradigms, it is clear that they are lacking in representation of understandings that stem from the Global South and from Indigenous knowledge.

Even in the United Nations Sendai Framework for Disaster Risk Reduction, there is an emphasis on Indigenous knowledge as being key to managing disasters in a way that is holistic. Ludwig (2017) highlights the areas where Indigenous knowledge picks up on potential risks when traditional scientific models fail to do so. Furthermore, the knowledge acquired is accumulated over a significant period of time, meaning that there is an innate understanding of the nature of things, especially related to the environment (Hiwasaki, Luna, Syamsidik, & Shaw, 2014). For example, flooding is a risk that has been addressed within Indigenous knowledge, due to the regularity of its occurrence and the cascading of events that stem from (Balay-As, Marlowe, & Gaillard, 2018). Much criticism has been posited at the binaries created in separating Indigenous understandings of risk from "western" science. Balay-As et al. (2018) have argued that this difference has resulted in the disenfranchisement of Indigenous knowledge, as scientific notions are still credited with being the most utilised, and rational. Much work in ethnography has sought to bridge the gap, while being aware of the dangers of exploiting Indigenous knowledge for its own purpose, as has been done in many instances through the problematic framing of 'integration' (see, for example, Mercer, Kelman, Taranis, and Suchet-Pearson (2010)) (a word that carries its own negative connotations in colonised and migratory communities). The canon of risk and disaster management literature is riddled with white, Western epistemologies that seek to retain and maintain its hegemonic position within the scholarship. Scholars such as Bankoff (2019) argue that the ideas of 'vulnerability' and 'safety' are heavily constructed from an ethnocentric position, that it produces a world that is reflective of a certain perspective to the omission of others. The decolonisation literature in risk stems back to the 1970s when discussing peoples' science (Wisner, O'Keefe, & Westgate, 1977), and in recent times this has developed to not only reflect a diverse understanding and acknowledgment of knowledge but also include this in risk assessment practice. Risk assessment practice in the Western canon has been predominantly based on technocratic approaches that are heavily reliant on quantitative systems and scientific underpinnings. Indigenous systems of knowledge tend to emphasise the culture-based knowledge systems that tend to be more participatory in nature (e.g. Allen, 2006; Fiddian-Qasmiyeh, 2019) and as such, more research should be undertaken to reflect these risk understandings.

Masculine Framings of Water and Risk

Similarly, masculine framings reign supreme. As the overarching technocratic narrative represents one of 'control', such as rerouting waterways and 'creating' new water for use through updated treatment technologies, one cannot ignore the undercurrents of hyper-masculine identities of power pervading this field (Vera Delgado & Zwarteveen, 2017). For too long, the feminist input into water engineering has relied predominantly on increasing the representation of women in water management, with little understanding of how the underlying structure is intrinsically masculine (Zwarteveen, 2008) – therefore affecting risk notions. An example is the endearing Smart Water Management approach, which essentially acts as an extension of this existing symbolism and identity of 'hegemonic masculinity' (Connell & Messerschmidt, 2005) pervasive in water engineering. Autocratic power as a feature of such a system reveals the ongoing struggle to harness and control natural environments, with hegemonic and nostalgic notions of modernisation still firmly intact. Power is the key motif in such environments. What can traditional notions of risk management offer feminism apart from continuing an age-old paradigm of perceived rationality through technocratic, power-hungry means? Very little, I would argue.

Instead, I ask whether there is a greater role for feminist rhetoric to play in restructuring the relationships we have with water, and also questioning the structural power that we, as water managers, try to impose on the resource.

Masculinity is well acknowledged to be an obsessively technology-focused identity (Lohan & Faulkner, 2004) that disproportionately influences water management and steers it towards the power-hungry motifs that serve to reinforce hegemonic gender norms and ideals. The preoccupation with infrastructure and technological advancements goes to the heart of Beck's (1992) notion of the risk society, creating solutions for our own self-imposed risks, while in turn generating new risks through this process. This is true of projects that introduce inter-basin transfers, of desalination and of other areas of technocratic decision-making that favour infrastructure solutions to solve the wicked problems of today, in turn creating their own social and environmental problems. Could the imposed structural masculinity in water management be creating larger risks than those that we are attempting to mitigate? These new narratives are not about taking the masculine preoccupation with 'control', but rather attempting to remove the power dynamic to work together with a resource that is essential to our existence. Feminist and post-colonial framings therefore can provide a valuable insight into critiquing current models of water governance to help us question the overreliance on technology-driven approaches to managing water, or whether there may be opportunities for social or fringe solutions to be incorporated into the water management mix. The embracing of varied perspectives provide the impetus needed to reconfigure our relationship to water, away from one that is paternalistic to an approach devoid of control-based obsession that works ultimately to serve the environment and, in turn, us.

Note

1 Refer to Hemming et al. (2018) for templates that can be used for the IDEA framework.

References

Adams-Hosking, C., McBride, M. F., Baxter, G., Burgman, M., De Villiers, D., Kavanagh, R., … McAlpine, C. A. (2016). Use of expert knowledge to elicit population trends for the koala (*Phascolarctos cinereus*). *Diversity and Distributions, 22*(3), 249–262. https://doi.org/10.1111/ddi.12400

All European Academies (ALLEA). (2018). *Loss of trust? Loss of trustworthiness? Truth and expertise today*. Retrieved from https://www.allea.org/wp-content/uploads/2018/06/ALLEA_Discussion_Paper_1_Truth_and_Expertise_Today-digital.pdf

Allen, K. M. (2006). Community-based disaster preparedness and climate adaptation: Local capacity-building in the Philippines. *Disasters, 30*(1), 81–101. https://doi.org/10.1111/j.1467-9523.2006.00308.x

Aven, T. (2014). *Risk, surprises and black swans: Fundamental ideas and concepts in risk assessment and risk management*. London: Routledge.

Aven, T. (2017). Improving risk characterisations in practical situations by highlighting knowledge aspects, with applications to risk matrices. *Reliability Engineering & System Safety, 167*, 42–48. https://doi.org/10.1016/j.ress.2017.05.006

Balay-As, M., Marlowe, J., & Gaillard, J. C. (2018). Deconstructing the binary between indigenous and scientific knowledge in disaster risk reduction: Approaches to high impact weather hazards. *International Journal of Disaster Risk Reduction, 30*, 18–24. https://doi.org/10.1016/j.ijdrr.2018.03.013

Bankoff, G. (2019). Remaking the world in our own image: Vulnerability, resilience and adaptation as historical discourses. *Disasters, 43*(2), 221–239. https://doi.org/10.1111/disa.12312

Beck, U. (1992). *Risk society: Towards a new modernity*. London: Sage Publications.

Beck, U. (2006). Living in the world risk society: A hobhouse memorial public lecture given on Wednesday 15 February 2006 at the London School of Economics. *Economy and Society, 35*(3), 329. Retrieved from https://ezp.lib.unimelb.edu.au/login?url=https://search.ebscohost.com/login.aspx?direct=true&db=edsgao&AN=edsgcl.152992829&site=eds-live&scope=site

Berman, J. M. (2020). *Anti-vaxxers: How to challenge a misinformed movement*. Cambridge, MA: MIT Press.

Bickerstaff, K., Lorenzoni, I., Pidgeon, N. F., Poortinga, W., & Simmons, P. (2008). Reframing nuclear power in the UK energy debate: Nuclear power, climate change mitigation and radioactive waste. *Public Understanding of Science, 17*(2), 145–169. https://doi.org/10.1177/0963662506066719

Bolsen, T., Palm, R., & Kingsland, J. T. (2019). The impact of message source on the effectiveness of communications about climate change. *Science Communication, 41*(4), 464–487. https://doi.org/10.1177/1075547019863154

Braman, D., Kahan, D. M., Slovic, P., & Gastil, J. (2006). Fear of democracy: A cultural evaluation of sunstein on risk. *Harvard Law Review, 119*, 1071.

Brewer, P. R., & Ley, B. L. (2013). Whose science do you believe? Explaining trust in sources of scientific information about the environment. *Science Communication, 35*(1), 115–137. https://doi.org/10.1177/1075547012441691

Burgman, M. (2016). *Trusting judgements: How to get the best out of experts*. Cambridge: Cambridge University Press.

Canadian Dental Association. (2010, October 28 2010). Ontario cities vote against water fluoridation. Retrieved from https://jcda.ca/article/a152

Carstairs, C. (2015). Debating water fluoridation before Dr. Strangelove. *American Journal of Public Health, 105*(8), 1559–1569. https://doi.org/10.2105/AJPH.2015.302660

Centers for Disease Control and Prevention. (1999). Achievements in Public Health, 1900–1999: Fluoridation of Drinking Water to Prevent Dental Caries. Retrieved from https://www.cdc.gov/mmwr/preview/mmwrhtml/mm4841a1.htm

CONCISE. (2020). *Communication role on perception and beliefs of EU citizens about science*. Retrieved from https://concise-h2020.eu/wp-content/uploads/2020/12/CONCISE_policy_brief_EN.pdf

Connell, R. W., & Messerschmidt, J. W. (2005). Hegemonic masculinity. *Gender & Society, 19*(6), 829–859. https://doi.org/10.1177/0891243205278639

Council of Standards Australia. (2009). *Risk management - principles and guidelines* (Vol. AS/NZS ISO 31000:2009). Sydney: Standards Australia.

Davies, A. (2006, 8th February 2006). Desalination plant dumped: It was a stinker with voters, to be frank. *Sydney Morning Herald*.

Douglas, M. (1992). *Risk and blame: Essays in cultural theory*. London: Routledge.

Drottz-Sjoberg, B.-M. (1991). *Perception of risk: Studies of risk attitudes, perceptions and definitions*. Stockholm: Center for Risk Research, Stockholm School of Economics.

Earle, T., C., & Cvetkovich, G. (2001). Social trust and culture in risk management. In G. Cvetkovich & R. E. Lofstedt (Eds.), *Social trust and the management of risk* (pp. 42–65). New York: Earthscane.

European Commission. (n.d.). Neonicotinoids. Retrieved from https://food.ec.europa.eu/plants/pesticides/approval-active-substances/renewal-approval/neonicotinoids_en

Facebook for Business. (2020). Fact-checking on Facebook: What publishers should know. Retrieved from https://www.facebook.com/business/help/182222309230722?helpref=uf_permalink

Fiddian-Qasmiyeh, E. (2019). Looking forward: Disasters at 40. *Disasters, 43*(S1), S36–S60. https://doi.org/10.1111/disa.12327

Fischhoff, B. (1995). Risk perception and communication unplugged: Twenty years of process. *Risk Analysis, 15*(2), 137–145. https://doi.org/10.1111/j.1539-6924.1995.tb00308.x

Fitchen, J. M. (1990). Cultural values affecting risk perception: individualism and the perception of toxicological risks. In L. A. Cox & P. F. Ricci (Eds.), *New risks: Issues and management* (pp. 599–607). Boston, MA: Springer US.

Forester, J. (1999). *The deliberative practitioner: Encouraging participatory planning processes*. Cambridge, MA: The MIT Press.

Freudenburg, W. R. (1988). Perceived risk, real risk: Social science and the art of probabilistic risk assessment. *Science, 242*(4875), 44–49. Retrieved from http://www.jstor.org/stable/1702491

Freudenburg, W. R. (2001). Risky thinking: facts, values and blind spots in societal decisions about risks. *Reliability Engineering & System Safety, 72*(2), 125–130. https://doi.org/10.1016/s0951-8320(01)00013-8

Ghosh, S. (2020). Twitter is adding fact-checking labels on tweets that link 5G with the coronavirus. *Business Insider Australia*. Retrieved from https://www.businessinsider.com.au/twitter-factchecks-tweets-5g-coronavirus-2020-6?r=US&IR=T

Giddens, A. (1991). *The consequences of modernity*. Palo Alto, CA: Stanford University Press.

Gillespie, E. (2020). #Sexysutton: How Victoria's Chief Health Officer reached cult status. *SBS The Feed*. Retrieved from https://www.sbs.com.au/news/the-feed/sexysutton-how-victoria-s-chief-health-officer-reached-cult-status

Giner-Sorolila, R., & Chaiken, S. (1997). Selective use of heuristic and systematic processing under defense motivation. *Personality and Social Psychology Bulletin, 23*(1), 84–97. https://doi.org/10.1177/0146167297231009

Groothuis, P. A., & Miller, G. (1997). The role of social distrust in risk-benefit analysis: A study of the siting of a hazardous waste disposal facility. *Journal of Risk and Uncertainty, 15*(3), 241–257. https://doi.org/10.1023/A:1007757326382

Hartley, K., & Kuecker, G. (2020). The moral hazards of smart water management. *Water International, 45*(6), 1–9. https://doi.org/10.1080/02508060.2020.1805579

Hemming, V., Burgman, M. A., Hanea, A. M., McBride, M. F., & Wintle, B. C. (2018). A practical guide to structured expert elicitation using the IDEA protocol. *Methods in Ecology and Evolution, 9*(1), 169–180. https://doi.org/10.1111/2041-210X.12857

Hiwasaki, L., Luna, E., Syamsidik, & Shaw, R. (2014). Process for integrating local and indigenous knowledge with science for hydro-meteorological disaster risk reduction and climate change adaptation in coastal and small island communities. *International Journal of Disaster Risk Reduction, 10*, 15–27. https://doi.org/10.1016/j.ijdrr.2014.07.007

Hmielowski, J. D., Feldman, L., Myers, T. A., Leiserowitz, A., & Maibach, E. (2013). An attack on science? Media use, trust in scientists, and perceptions of global warming. *Public Understanding of Science, 23*(7), 866–883. https://doi.org/10.1177/0963662513480091

Hurlimann, A., & Dolnicar, S. (2010). When public opposition defeats alternative water projects – the case of Toowoomba Australia. *Water Research, 44*(1), 287–297.

International Organization for Standardization. (2009a). *Risk management - risk assessment techniques* (IEC31010:2009). Retrieved from https://www.iso.org/standard/51073.html

International Organization for Standardization. (2019). *Risk management – Risk assessment techniques* (IEC31010:2019). https://www.iso.org/standard/72140.html

Johnson, T. J., & Kaye, B. K. (2004). Wag the Blog: How reliance on traditional media and the internet influence credibility perceptions of weblogs among blog users. *Journalism & Mass Communication Quarterly, 81*(3), 622–642.

Kahan, D. M., Braman, D., Slovic, P., Gastil, J., & Cohen, G. (2009). Cultural cognition of the risk and benefits of nano-technology. *Nature Nanotechnology, 4*(2), 87–91.

Kosovac, A., & Davidson, B. (2020). Is too much personal dread stifling alternative pathways to improving urban water security? *Journal of Environmental Management*, 265, 110496. https://doi.org/10.1016/j.jenvman.2020.110496

Kosovac, A., Davidson, B., & Malano, H. (2019). Are we objective? A study into the effectiveness of risk measurement in the water industry. *Sustainability, 11*(5), 1279. https://doi.org/10.3390/su11051279

Kosovac, A., Hartley, K., & Kuecker, G. (2021, 9 September). When water runs out of time there will be no tech solution. *Smart Water Magazine, 9*. Retrieved from https://smartwatermagazine.com/bimonthly/9

Kosovac, A., Hurlimann, A., & Davidson, B. (2017). Water experts' perception of risk for new and unfamiliar water projects. *Water (20734441), 9*(12), 1.

Kosovac, A., & McLeod Logan, T. (2021). Resilience: Lessons to be learned from safety and acceptable risk. *Journal of Safety Science and Resilience, 2*(4), 253–257. https://doi.org/10.1016/j.jnlssr.2021.10.002

Lewandowsky, S., Ecker, U. K. H., Seifert, C. M., Schwarz, N., & Cook, J. (2012). Misinformation and its correction. *Psychological Science in the Public Interest, 13*(3), 106–131. https://doi.org/10.1177/1529100612451018

Lohan, M., & Faulkner, W. (2004). Masculinities and technologies. *Men and Masculinities, 6*(4), 319–329. https://doi.org/10.1177/1097184x03260956

Ludwig, D. (2017). The objectivity of local knowledge. Lessons from ethnobiology. *Synthese, 194*(12), 4705–4720. https://doi.org/10.1007/s11229-016-1210-1

Lupton, D. (2013). *Risk* (2nd ed.). New York and Oxon: Routledge.

Ma, Y., Dixon, G., & Hmielowski, J. D. (2019). Psychological reactance from reading basic facts on climate change: The role of prior views and political identification. *Environmental Communication, 13*(1), 71–86. https://doi.org/10.1080/17524032.2018.1548369

Malka, A., Krosnick, J. A., & Langer, G. (2009). The association of knowledge with concern about global warming: Trusted information sources shape public thinking. *Risk Analysis, 29*(5), 633–647. https://doi.org/10.1111/j.1539-6924.2009.01220.x

Margolis, H. (1996). *Dealing with risk: Why the public and the experts disagree on environmental issues*. Chicago, IL: University of Chicago Press.

Mercer, J., Kelman, I., Taranis, L., & Suchet-Pearson, S. (2010). Framework for integrating indigenous and scientific knowledge for disaster risk reduction. *Disasters, 34*(1), 214–239. https://doi.org/10.1111/j.1467-7717.2009.01126.x

Mills, R. A., Karajeh, F., & Hultquist, R. H. (2004). California's task force evaluation of issues confronting water reuse. *Water Science and Technology: A Journal of the*

International Association on Water Pollution Research, 50(2), 301–308. Retrieved from https://ezp.lib.unimelb.edu.au/login?url=https://search.ebscohost.com/login.aspx?direct=true&db=mnh&AN=15344805&site=eds-live&scope=site
Myers, T. A., Kotcher, J., Stenhouse, N., Anderson, A. A., Maibach, E., Beall, L., & Leiserowitz, A. (2017). Predictors of trust in the general science and climate science research of US federal agencies. *Public Understanding of Science, 26*(7), 843–860. https://doi.org/10.1177/0963662516636040
Nisbet, E. C., Cooper, K. E., & Garrett, R. K. (2015). The Partisan Brain. *The ANNALS of the American Academy of Political and Social Science, 658*(1), 36–66. https://doi.org/10.1177/0002716214555474
Nisbet, M. C., & Myers, T. (2007). The polls trends: Twenty years of public opinion about global warming. *Public Opinion Quarterly, 71*(3), 444–470. https://doi.org/10.1093/poq/nfm031
Ochi, S. (2021). 'Life communication' after the 2011 Fukushima nuclear disaster: What experts need to learn from residential non-scientific rationality. *Journal of Radiation Research, 62*(Supplement_1), i88–i94. https://doi.org/10.1093/jrr/rraa135
Pidgeon, N., Slovic, P., Kasperson, R. E (2003). The Social Amplification of Risk. Cambridge: Cambridge University Press.
Po, M., & Nancarrow, B. E. (2004). *Literature Review: Consumer perceptions of the use of reclaimed water for horticultural irrigation*. Retrieved from Perth CSIRO Land and Water
Sanz-Menendez, L., & Cruz-Castro, L. (2019). The credibility of scientific communication sources regarding climate change: A population-based survey experiment. *Public Understanding of Science, 28*(5), 534–553. https://doi.org/10.1177/0963662519840946
Sewell, W. R. D. (1971). Behavioral responses to changing environmental quality. *Environment and Behavior, 3*(2), 119–122. https://doi.org/10.1177/001391657100300201
Siegrist, M. (2021). Trust and risk perception: A critical review of the literature. *Risk Analysis, 41*(3), 480–490. https://doi.org/10.1111/risa.13325
Siegrist, M., & Cvetkovich, G. (2000). Perception of hazards: The role of social trust and knowledge. *Risk Analysis, 20*(5), 713–719. https://doi.org/10.1111/0272-4332.205064
Siegrist, M., Cvetkovich, G., & Roth, C. (2000). Salient value similarity, social trust, and risk/benefit perception. *Risk Analysis, 20*(3), 353–362. https://doi.org/10.1111/0272-4332.203034
Siegrist, M., Gutscher, H., & Earle, T., C. (2005). Perception of risk: The influence of general trust, and general confidence. *Journal of Risk Research, 8*(2), 145–156. https://doi.org/10.1080/1366987032000105315
Sivagurunathan, V., Kosovac, A., & Khan, S. (2022). Urban potable reuse: Contrasting perspectives of water industry professionals and elected politicians in Sydney, Australia. *Water International, 47*(1), 73–91. https://doi.org/10.1080/02508060.2021.1969768
Sjoberg, L., & Sjoberg-Drottz, B. (2008). Risk perception by politicians and the public. *Energy & Environment, 19*(3–4), 455–482.
Slovic, P. (1990). Perception of risk from radiation. In W. K. Sinclair (Ed.), *Proceedings of the twenty-fifth annual meeting of the national council on radiation protection and measurements* (Vol. 11, pp. 73–87). Bethesda, MD: NCRP.
Slovic, P. (1992). Perception of risk. In S. Krimsky & D. Golding (Eds.), *Social theories of risk* (pp. 117–153). Westport, CT: Praeger Publisher.
Slovic, P. (1993). Perceived risk, trust, and democracy. *Risk Analysis: An International Journal, 13*(6), 675–682.

Slovic, P., Finucane, M. L., Peters, E., & MacGregor, D. G. (2004). Risk as analysis and risk as feelings: Some thoughts about affect, reason, risk, and rationality. *Risk Analysis, 24*(2), 311–322. https://doi.org/10.1111/j.0272-4332.2004.00433.x

Sofoulis, Z. (2005). Big water, everyday water: A sociotechnical perspective. *Continuum, 19*(4), 445–463. https://doi.org/10.1080/10304310500322685

Sunstein, C. R. (2005). *Laws of fear: Beyond the precautionary principle*. Cambridge: Cambridge University Press.

Taleb, N. N. (2008). *The Black Swan: The impact of the highly improbable*. London: Penguin Books Limited.

Tillman, R. (2021, July 16). Olivia Rodrigo, Dr. Anthony Fauci star in White House vaccination video. *Spectrum News*. Retrieved from https://spectrumlocalnews.com/nys/central-ny/news/2021/07/16/olivia-rodrigo-anthony-fauci-covid-vaccination-video

Tversky, A., & Kahneman, D. (1973). *Judgment under uncertainty: Heuristics and biases*. Oregon Research Institute, Springfield.

Vera Delgado, J. R., & Zwarteveen, M. (2017). Queering engineers? Using history to re-think the associations between masculinity and irrigation engineering in Peru. *Engineering Studies, 9*(2), 140–160. https://doi.org/10.1080/19378629.2017.1361427

Wasko, C., Nathan, R., Stein, L., & O'Shea, D. (2021). Evidence of shorter more extreme rainfalls and increased flood variability under climate change. *Journal of Hydrology, 603*, 126994. https://doi.org/10.1016/j.jhydrol.2021.126994

Wasko, C., Sharma, A., & Johnson, F. (2015). Does storm duration modulate the extreme precipitation-temperature scaling relationship? *Geophysical Research Letters, 42*(20), 8783–8790. https://doi.org/10.1002/2015GL066274

Wisner, B., O'Keefe, P., & Westgate, K. (1977). Global systems and local disasters: The untapped power of peoples' science. *Disasters, 1*(1), 47–57. https://doi.org/10.1111/j.1467-7717.1977.tb00008.x

Woods, M. (2020, March 3). Canada's Health Officials are the heroes we need during this Coronavirus pandemic. *The Huffington Post*. Retrieved from https://www.huffingtonpost.ca/entry/coronavirus-canada-public-health-official_ca_5e753865c5b6f5b7c5444f71

Zinn, J. (2008a). Heading into the unknown: Everyday strategies for managing risk and uncertainty. *Health, Risk and Society, 10*(5), 439–450.

Zinn, J. (2008b). *Social theories of risk and uncertainty: An introduction*. Malden, MA: Blackwell Publishing.

Zwarteveen, M. (2008). Men, masculinities and water powers in irrigation. *Water Alternatives, 1*(1), 111–130.

7 Conclusion

Risk as Bigger than the Matrix

Decision-making, especially related to risk, is an inherently subjective process. Some may argue that risk *can* be objective, on the basis of using data that is irrefutable. However, the decisions that hinge on such data rarely exist separately from the socio-cultural, especially in environmental decision-making. Whether it be upgrading a pipe, changing water supplies or imposing water restrictions, these decisions cannot be based on the purely technical. These choices will impact people; they will impact flora, fauna and social systems. They will be political. The decisions as such are guided by the pull of various areas that cannot be measured objectively.

If only statistical information is utilised for decision-making, then this can result in the disregarding of many social outcomes or consequences not considered by water practitioners, and this could also fuel public disenchantment and ultimately lead to loss of public trust towards the water authority/utility (Healy, 2001).

For example, the water restrictions of 2008 in the South East of Australia were imposed as a result of years of consecutive droughts (titled the 'Millennium Drought') that were predicted would continue into the future by climate experts (Bureau of Meteorology, 2015). In order to safeguard water supply, water experts had conducted balance modelling to determine the levels of restrictions required to see Melbourne through the predicted drought. Water restrictions were subsequently imposed on the community at a state policy level, which incorporated no irrigation of sports fields and urban open spaces (Melbourne Water, ND). At a basin level, this was required to reduce demand when purely considering water balance models. By limiting the water being used, there can be an added water security buffer during the summer when there is little replenishment of the dams during the year. The impact of the water restrictions was profound, but not in the way that water experts predicted. Due to the drying of sports fields and green spaces, there was a reduction in social capital within the community as Australian football matches, picnics and other outdoor events where people would often gather and socialise would cease (Weller & English, 2008). This subsequently led to

DOI: 10.4324/9781003432647-7

detrimental mental health impacts on many parts of the community, as they had lost their social connection touch points (Weller & English, 2008). In this circumstance, the sole reliance on data modelling did not predict a number of adverse social outcomes (and new risks) from the policy action. Input from the community in decision-making in this instance was paramount (Syme, 2008). A way of doing this is to incorporate a co-determination process that is central to achieving a type of decision-making that integrates public values together with technical expertise (Renn, 2001, 2006). The nature of our society is pluralist, incorporating differing value systems and worldviews across the spectrum, thus ensuring that it may be difficult to conduct truly representative collective decision-making (Renn, 2006). As Braman, Kahan, Slovic and Gastil (2006) attest, "[b]ecause citizens' fears express their cultural visions of how society should be organised, the line between 'considered values' and 'irrational fears' often proves illusory". The nature of the fears expressed within the community and the facts they refute display expressions of their values of how they see the world, a critical component to understanding risk perception.

As a result, understanding the psycho-social theories within this book is a first step. It is about recognising that risk is formed from how easily a person imagines (and feels about) a scenario, how they believe society should be structured (each manages their own risk vs. government-managed) or how the options interact with personal (or organisational) values.

Overall, the main argument to take away from this book is that risk cannot be reduced to a simple matrix. That we need to question how and why we make the decisions we do, and critique (in a healthy way) decisions being made around us. Risk processes today are worryingly not fit for purpose in today's uncertain world. The over-quantification to produce a seemingly objective result has had its day, and it is time to understand other conceptualisations of risk. The sociological and psychological approaches to risk highlight compelling ways of understanding risk and the role of the risk assessor. Although many would be loathed to do away with the risk matrix entirely, there is much scope to reflect upon how biases can creep into assessments.

Who Should Make Decisions?

I remember standing in a work kitchenette one day, discussing some of the public concerns relating to an upcoming major project that was happening. A project manager[1] chimed in, saying that they were having a lot of trouble being able to acquire properties to construct their large pipeline. People were, somewhat unsurprisingly, refusing to give up their homes. The project manager was exasperated, but also understood the concerns. There was the difficult tension between installing the pipeline (to increase the city's water security for the next 50 years) and a person's right to stay in the home they have developed a strong attachment to. One of the more experienced Engineers walked past at this point in the conversation and reminisced about the 'glory days', i.e. the days where there was a large and all-powerful Board of Works who would do what needed to be done. "There was none of this negotiation nonsense. If you need that property for the good of everyone else, you take it, no negotiations or questions asked. Nowadays, there is too much pandering to the whim of every person".

Between those days to now, there have been large shifts in socio-political relations, especially those associated with how a citizen relates to their own government. Opinion polls can serve to rid ministers of jobs alarmingly quickly, and the water practitioners at the project level will often feel the pressure to ensure public satisfaction.

As discussed in Chapter 5, the types of projects that water practitioners decide upon are sensitive to public relations factors, as risk to reputation features heavily in concerns regarding risk (Kosovac & Davidson, 2020). Experts within water authorities have concerns regarding what this public risk perception is, especially if it is likely to differ from their own. What has changed, then, is that water practitioners are asked to do more today than they ever have in the past. Their consideration of laypeople's views on projects features heavily in this work, and is also reflected in their risk appraisals. This is a heartening move towards incorporating others' perspectives and a (albeit slow) move away from technocracy.

The implementation of policy surrounding issues of public collective action is highly reliant on risk assessments of both experts and laypeople to inform decision-making. This is not to say that experts purport to hold all relevant knowledge or are purely rational decision-makers. As discussed, experts carry their own psychological biases and values which inform their stances, while in much the same way relevant local knowledge can be paramount to supplement existing expertise. Although the rhetoric regarding public disdain for scientists has been rife over the last three decades (see Slovic, 1992 for similar arguments against experts that we see today in Nichols, 2017), the evidence shows that the general public is largely trusting of experts. Controversial issues such as the implementation of 5G technology or the construction of nuclear power facilities present key flash point debates where public sentiment has in some instances diverged from expert opinion. But these examples do not highlight a widespread distrust of experts, but rather the effectiveness of democratic practices in steering public policy.

In creating this synthesis of the fields of risk analysis, expertise, trust and democratic practice, it is apparent that there is no 'war' that is being waged between laypeople and experts but rather a constant co-existence that, at times, carries its own healthy tensions. The argument within this book does not presuppose that experts are the only legitimate sources of information, nor should their advice act as the only input in key public sector decisions. As has been discussed, the community more broadly provides crucial insight into social matters as well as into value prepositions of policies of public interest. As a society we want to discuss risks, we want to understand risks and we want to have a say in the risks that we are exposed to. Without experts and the facts they provide, we can never know the true extent of these risks (if a 'real' risk is one's ontological basis) nor can we present a balanced, informed viewpoint. Unfortunately, as previous studies have shown, people demand more risk mitigation by the government than they do themselves (Sjöberg, 2001). It is for this reason that governments are often heavily scrutinised in terms of risk-based decision-making and also why water practitioners need to be increasingly cognisant of their biases and the role of risk in our society.

Note

1 Note that I use the words 'water practitioners', 'project managers' and 'experts' interchangeably within this book. This also includes 'Engineer'. Water Practitioners are often Engineering graduates.

References

Braman, D., Kahan, D. M., Slovic, P., & Gastil, J. (2006). Fear of democracy: A cultural evaluation of Sunstein on risk. *Harvard Law Review, 119*, 1071.

Bureau of Meteorology. (2015). Recent rainfall, drought and southern Australia's long-term rainfall declining. Retrieved from http://www.bom.gov.au/climate/updates/articles/a010-southern-rainfall-decline.shtml

Healy, S. (2001). Risk as social process: The end of 'the age of appealing to the facts'? *Journal of Hazardous Materials, 86*(1–3), 39–53. https://doi.org/10.1016/s0304-3894(01)00254-0

Kosovac, A., & Davidson, B. (2020). Is too much personal dread stifling alternative pathways to improving urban water security? *Journal of Environmental Management, 265*, 110496. https://doi.org/10.1016/j.jenvman.2020.110496.

Melbourne Water. (n.d.). 2002–2009 Drought and water restriction hit hard. Retrieved from http://www.waterstory.melbournewater.com.au/content/our-urban-water-story/drought-and-water-restrictions-hit-hard

Nichols, T. M. (2017). *The death of expertise: The campaign against established knowledge and why it matters*. New York: Oxford University Press.

Renn, O. (2001). The need for integration: Risk policies require the input from experts, stakeholders and the public at large. *Reliability Engineering & System Safety, 72*(2), 131–135. https://doi.org/10.1016/s0951-8320(01)00014-x

Renn, O. (2006). Participatory processes for designing environmental policies. *Land Use Policy, 23*(1), 34–43. https://doi.org/10.1016/j.landusepol.2004.08.005

Sjöberg, L. (2001). Author's reply: Whose risk perception should influence decisions? *Reliability Engineering & System Safety, 72*(2), 149–151. https://doi.org/10.1016/s0951-8320(01)00016-3

Slovic, P. (1992). Perception of risk. In S. Krimsky & D. Golding (Eds.), *Social theories of risk* (pp. 117–153). Westport, CT: Praeger Publisher.

Syme, G. (2008). The social and cultural aspects of sustainable water use. In L. Crase (Ed.), *Water policy in Australia: The impact of change and uncertainty* (pp. 231–243). Washington, DC: Resources for the Future.

Weller, S., & English, A. (2008). *Socioeconomic impact study of water restrictions on turf sports grounds in Metropolitan Melbourne*. Retrieved from Melbourne.

Index

Note: **Bold** page numbers refer to tables; *Italic* page numbers refer to figures and page numbers followed by "n" denote endnotes.

Printed in the United States
by Baker & Taylor Publisher Services